The *Obelus* Set Theory of Equity Distribution

AuthorHouse™
1663 Liberty Drive
Bloomington, IN 47403
www.authorhouse.com
Phone: 1-800-839-8640

First published by AuthorHouse   09/13/2011

ISBN: 978-1-4567-8725-7 (sc)

Printed in the United States of America

Any people depicted in stock imagery provided by Thinkstock are models, and such images are being used for illustrative purposes only.
Certain stock imagery © Thinkstock.

This book is printed on acid-free paper.

Cover page design by Jonnyall

The *OBELUS* Set Theory
of Equity Distributions

Jonathan Yalley

Authorhouse UK Ltd
500 Avebury Boulevard
Milton Keynes
MK9 2BE
United Kingdom

Quotes on Sharing

● "Share everything. Don't take things that aren't yours. Put things back where you found them."   *Robert Fulghum*
{Equity Distribution}

● "The greatest gift is to give people your enlightenment, to share it. It has to be the greatest."   *Buddha*
{Unselfish Distributor}

● "There is no delight in owning anything unshared."   *Seneca*
{Propensity}

● "Life is about giving, and the rest is taken care of."   *Peter Cajander*
{Unit Donor Principle}

● "The miracle is this: the more we share the more we have."   *Leonard Nimov*
{Set Resonance}

● "All for one and one for all"   *Proverb quotes*
{Standing Set Surrogate Resonance}

● "We must ensure that the global market is embedded in broadly shared values and practices that reflect global social needs, and that all the world's people share the benefits of globalization."
{Wholistic Distributive Power Set}

Kofi Annan

© UNESCO Tom Fleishhauer/2001.

● "When you dream alone, with your eyes shut, asleep, that dream is an illusion. But when we dream together, sharing the same dream, awake and with our eyes wide open, then that dream becomes reality!"

Unknown

● "What is absolutely true is always correct, everywhere, all the time, under any condition. An entity's ability to discern these things is irrelevant to that state of truth."

Steven Robiner

v

To my dear mother Mrs. Rose Anna Yalley (nee Arthur), the rest of my family, relatives and friends.

PREFACE

In mathematics, distributions which are also called generalized functions are objects which generalize functions and probability distributions. However, the distributive aspects of division processes investigated in this book are different from those of distribution theory.

The operation of arithmetic commonly called division basically entails the separation of object(s) into two or more equity groups or sets. It can practically be related to the act of distributing entities among individual(s) on equity basis. In general, the process of division is characterized by an interactive phase which leads to an evenhanded distributive phase. The mathematical basis underlining this process of sharing, which is an equity distributive process, will be referred to as the '*Obelus* Set Theory'. The word '*obelus*' is a Greek word for dagger and is generally used to represent the division sign, $\div$. The *Obelus* set theory is a ubiquitous mathematical encapsulization of the process of division. It sets forth a paradigm shift in the logics and symbolic representations of mathematical division.

Since mathematics is not just about numbers but also about its meaning to reality, the provision of usable formulations that has correspondence to or at least an approximation to reality is an important need for any meaningful scientific endeavour. Thus, when division is analyzed qualitatively and quantitatively as a process of equitable distribution, the process naturally becomes congenially realistic and its end result is physically meaningful. This has the advantage of avoiding arithmetic conundrum such as infinity, undefined and indeterminate forms.

The formalization of the mathematics of distribution sets that is presented in this book has resulted from pragmatic and more so methodological motivations rather than one of a priori origin. Since the interrelationships that exist within the material world are distributional in nature, there seems reason to posit the process of division as universal. It represents the abstraction of the interrelationship between the real world and the world of mathematical abstraction. In general, some of the concepts to be encountered in this book include surrogation, propensity, redistribution, resonance, zeroth number field, mapping statistics, continuum hypothesis, geometrical point and the all important computer science P versus NP problem.

I take this opportunity to render my profound gratitude to Dr. Clifton Keller, Louisville, Kentucky, USA for his invaluable encouragement through his candid comment on the originality of the key concepts of the manuscript. Certainly, my special appreciation goes to my family members for their myriad support. I also would like to say thank you to the publisher, AuthorHouse for their invaluable service. Finally and most importantly, I give eternal thanks to God for everything.

Jonathan Yalley
October 12, 2010.

CONTENTS

CHAPTER 1

DIAGNOSIS OF DIVISION PROCESS

To begin with, arithmetic division will be considered as a process involving three sets or groups namely, the entity meant for distribution called '**distributum**', the recipient(s) called '**distribient**' and finally the **distributor** who oversees the division process. For example, if six oranges are to be shared between two people it is represented in contemporary arithmetic as: 7/2. Here, the number 7 is the distributum, the number 2 the distribient and the bar can be used to represent the distributor. In such a practical scenario, the distributor is always the holder of the distributum before the sharing process and the holder of the remainder after the sharing process. The result of the ratio 7/2 is 3½ called the quotient. Notice that the value ½ is implying another division scenario. In terms of whole number distribution, the result of the ratio 7/2 can be expressed as, 3 remainder 1. The number 3 is referred to as the '**quota**' while the number 1 is the **remainder**. Hence, the sum of the 'quota' and the remainder must be equal to the 'distributum'. The distributive power set, will be the foundation of this general division analysis. The key here is to keep track of who or where each portion of the distributum goes. This allows for a wholistic account of the distribution which is paramount for the realization of a down-to-earth interpretation of the arithmetic operation of division.

BASIC CASE STUDY ANALYSIS OF SHARING

The search for a complete overview of the interactive world of distribution may sound difficult and esoteric to accomplish. However the quantity, quality and variety of information that is realistically available in a sharing interaction is not beyond measure if

Case Study A: *The unselfish mood.*

Figure 1. Illustration of an unselfish distribution.

and only if all the processes involved are captured and analyzed in a systematic and equitable fashion. To better understand the important ideas underlying various sharing situations, the following case studies are used to make crystal clear each scenario.

Here, a teacher shares 6 oranges equally among 3 students as illustrated in figure 1. In terms of pictorial arithmetic, the above equity distribution scenario on the basis of a complete overview can be depicted as shown below in figure 2.

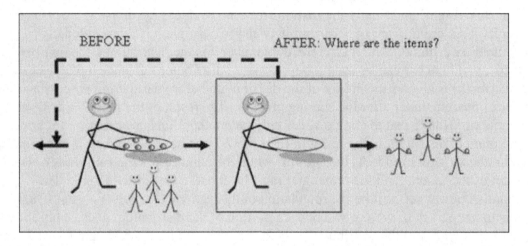

Figure 2. A pictorial depiction of the arithmetic process in an unselfish distribution.

Notice the additional pictorial data concerning 3 recipients during the before-stage of figure 2 scenario in contrast with that of figure 1. The transformation of the pictorial arithmetic scenario of figure 2 to symbolic expression must be structured such that it is logical and orderly. With this in mind, a full fledge numerical quasi-arithmetic representation will be,

$$[0]\frac{6}{3} = 2$$

where the retentive phase in the sharing interaction is symbolized by the square bracket, [ ]. The 0 in the square bracket represents the amount left in the teacher's (i.e. distributor's) hand prior to the distribution, the 6 (i.e. dividend or distributum) is the amount of oranges intended for equal distribution, the 3 (i.e. divisor) is the number of students who are the recipients and the 2 after the equal sign (i.e. quotient) is the number of orange(s) each student gets.

Case Study B: *The selfish mood.*

Here, teacher is unwilling to share the 6 oranges among any of the 3 students (or among any student(s)). This situation is depicted in figure 3 below.

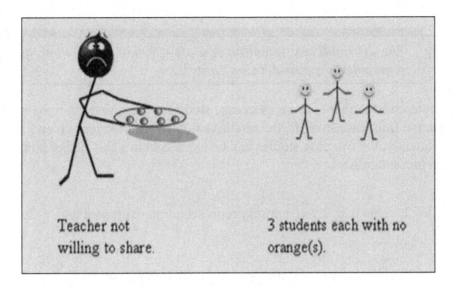

Figure 3. Illustration of a selfish distribution.

The corresponding pictorial arithmetic representation of the unwilling scenario in figure 3 is depicted in figure 4 below.

Figure 4. A pictorial depiction of the arithmetic process in a selfish distribution.

From the scenario under consideration, the corresponding numerical quasi-arithmetic representation will be: $[6]\dfrac{6}{0} = 0$ where 6 in the square bracket (i.e. retention phase) represents the amount left in the teacher's (i.e. distributor's) hand prior to the distribution, the 6 (i.e. dividend or distributum) is the number of oranges not intended for equal distribution, the 0 (i.e. divisor) is the number of student(s) who are recipient(s) (i.e. distribient) and the 0 after the equal sign (i.e. quotient) is the number of orange(s) each student gets. Generally, the number in the bracket called the '**retention bracket**' represents the remainder of the equity distribution process. It is different from a remainder which is what remains after division takes place. By definition,

Hence, the retentive number of oranges in case study A is zero and the remainder is zero too. In contrast, from case study B, the retentive number of oranges is 6 and its remainder is 0. Consequently, the two case studies can be expressed in a full fledge numerical arithmetic representation as,

i. $[0] + \left\langle \dfrac{6}{3} \right\rangle = 2$. This is by contemporary representation expressed as $\dfrac{6}{3} = 2$.

ii. $[6] + \left\langle \dfrac{6}{0} \right\rangle = 0$. This is by contemporary representation expressed as $\dfrac{6}{0} = undefined$.

Contemporary arithmetic does not in general allow division by zero. [1] The distribution density is denoted by $\langle \ \rangle$ and the divisor is representative of the distribution density.

Observe here that both wholistic symbolic arithmetic expressions, (i) and (ii) for Case Studies A and B are in contrast to their respective contemporary expressions, are incomparable in terms of quantity, quality and variety of information derived.

Finding the Effective Quota

From the above expressions, if division is truly an inverse of multiplication, then the RHS (right hand side) must be equal to the LHS (left hand side) for each expression. Thus for (i), multiplying through by 3 gives

$$[0] \times 3 + \left\langle \dfrac{6}{3} \right\rangle \times 3 = 2 \times 3.$$

Using BODMAS as the order of precedence of arithmetic operations to evaluate results in

$$0 + 6 = 6$$

which implies that

$$\text{LHS} = \text{RHS} = 6.$$

The result 6 represents the '**effective distribution quota**', **EDQ**. The 6 therefore means 6 out of 6 oranges were distributed to 3 students. For (ii), multiplying through by 0 gives

$$[6] \times 0 + \left\langle \dfrac{6}{0} \right\rangle \times 0 = 0 \times 0$$

which boils down to $0 + 0 = 0$. That is

$$\text{LHS} = \text{RHS} = 0.$$

The number 0 represents the effective distribution quota. It implies 0 out of the 6 oranges was distributed to nobody which is true.

The Zero Numerator and Zero Denominator Divisions Equivalence

Generally, in mathematics any integer divided by zero (a case of zero denominator) is said to be undefined. Thus, 6 divided by 0 until now will be undefined. However, the realistic approach introduced here, defines it as equal to zero. Let us see what happens in the case of a zero numerator. In the case of sharing zero oranges to say 6 students figure 5 shows the pictorial arithmetic representation.

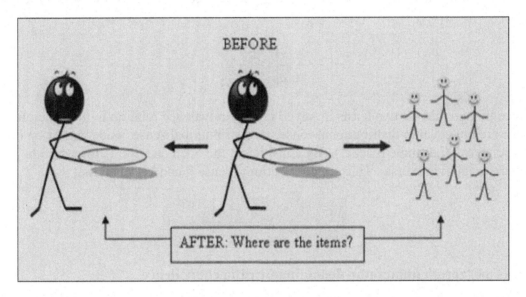

Figure 5. A pictorial depiction of the arithmetic process in the unselfish distribution of nothing.

The corresponding numerical quasi-arithmetic representation for the above scenario is

$$[0]\frac{0}{6} = 0.$$

Expanding results in a full fledge numerical arithmetic representation expressed as

(iii)
$$[0] + \left\langle \frac{0}{6} \right\rangle = 0$$

Multiplying through by 6 gives

$$([0] \times 6) + \left(\left\langle \frac{0}{6} \right\rangle \right) 6 = (0) \times 6$$
$$0 + (0)6 = 0$$
$$0 + 0 = 0$$
$$0 = 0.$$

5

The zero after the equal sign represents the effective distribution quota. From the last equation, the implication derived is that zero out of zero oranges was distributed to 6 students. This is the distributional sense. Hence, comparatively the '**effective distribution quota**' is

$$\left|\frac{6}{0}\right| \equiv \left|\frac{0}{6}\right| = 0.$$

This is the paradoxical statement between division by zero and a zero divided by a non-zero.

Generally, if x is a variable then the zeroth numerator-denominator ratio equality is given by

$$\left|\frac{x}{0}\right| \equiv \left|\frac{0}{x}\right| = 0$$

Notice though that the two terms involved in the equivalence relation in the generality state above are not equal in distributional sense. A distributional sense is denoted by an open or close chevron (i.e. angle bracket). By conversion, the open chevron is for a zeroth-numerator and vice versa. Thus, in distributional sense it can be stated that

$$\left\langle \frac{x}{0} \neq \frac{0}{x} \right\rangle \quad but \quad \left|\left\langle \frac{x}{0} \right| \equiv \left|\frac{0}{x} \right\rangle \right| = 0$$

This is the '**zeroth numerator-denominator ratio equivalence**'.

As already shown, by definition

> *The distributional sense is given by the sum of the ratio of retention to recipients plus dividend.*

On the other hand,

> *The distribution density is given by the ratio of the dividend to the divisor which is equal to the sum of the quotient minus retention.*

From (i), (ii) and (iii) the '**distribution densities**' are respectively re-stated as

i. $\left\langle \frac{6}{3} \right\rangle = 2 - [0] = 2$ ii. $\left\langle \frac{6}{0} \right\rangle = 0 - [6] = -6$ iii. $\left\langle \frac{0}{6} \right\rangle = 0 - [0] = 0$

The negative 6 found in (ii) means that each of the 6 oranges is taken away from the 3 students technically leaving the said 3 students with nothing. Note that the open and close chevrons are used around the ratio to indicate distribution density. Also, the distribution densities in (ii) and (iii) are not equal. That in (ii) is less than that in (iii). However, in both (ii) and (iii), both recipient(s) technically gained 'nothing'. This situation was earlier on referred to as the '**paradoxical equivalence between division by zero and a zero division**'. A concise explanation and proof of the nothingness of dividends in both conditions (ii) and (iii) will be given under '*The Concept of Set Resonance*' and the '*Proof of Division by Zero Solution*' under '*The Logarithm of Division by Zero*'

Generally, it can be stated that

> *The result of a distribution density, say* $\left\langle \dfrac{a}{b} \right\rangle$ *is a representation of the*
>
> 1. *Absolute retention if it is a negative value.*
>
> 2. *Quotient and/or remainder if it is a positive value.*

THE CONCEPT OF RECIPROCITY

This technique is used to attain a non-zero dividend in a sharing situation. Can the reciprocity between division and multiplication be proven here? Of course yes. To do this, take the 'double distribution density inverse' of both numerator and denominator. That is, in general

$$\left\langle \frac{\text{numerator}}{\text{denominator}} \right\rangle = \left\langle \frac{\left\langle \dfrac{\text{numerator}}{1} \right\rangle}{\left\langle \dfrac{\text{denominator}}{1} \right\rangle} \right\rangle$$

So considering the case in (i): $\left\langle \dfrac{6}{3} \right\rangle = 2$.

Applying the concept of reciprocity, we get

$$\left\langle \frac{\left\langle \dfrac{6}{1} \right\rangle}{\left\langle \dfrac{3}{1} \right\rangle} \right\rangle = 2.$$

By cross multiplying we get

$$\left\langle \frac{6}{1} \right\rangle = 2 \left\langle \frac{3}{1} \right\rangle.$$

This can be written as 6 = 6 and which is true and valid.

7

For the case in (ii) (see (ii) for proof),

$$\left\langle \frac{6}{0} \right\rangle = -6$$

Applying the concept of reciprocity, we get

$$\left\langle \frac{\left\langle \frac{6}{1} \right\rangle}{\left\langle \frac{0}{1} \right\rangle} \right\rangle = -6.$$

By cross multiplying we obtain

$$\left\langle \frac{6}{1} \right\rangle = -6 \left\langle \frac{0}{1} \right\rangle.$$

Notice here that the RHS cannot have a zero dividend. The fact is that something must be shared and this is attained through reciprocity. Hence,

$$RHS = \left\langle \frac{-6}{\left\langle \frac{1}{0} \right\rangle} \right\rangle.$$

This is a form of second degree division. However, finding the distribution density of both sides (i.e. LHS and RHS) gives the following. For the LHS,

$$\left\langle \frac{6}{1} \right\rangle = 6$$

and for the RHS

$$\left\langle \frac{-6}{\left\langle \frac{1}{0} \right\rangle} \right\rangle.$$

But

$$\left\langle \frac{1}{0} \right\rangle = -1.$$

Therefore,

$$\left\langle \frac{-6}{\left\langle \frac{1}{0} \right\rangle} \right\rangle = \left\langle \frac{-6}{-1} \right\rangle = \left\langle \frac{6}{1} \right\rangle = 6.$$

Thus, LHS = RHS = 6, which is true and valid. Notice the importance of the distribution density. It has helped to solve an impossible arithmetic that has existed for as long as the existence of arithmetic itself.

For the case in (iii):

$$\left\langle \frac{0}{6} \right\rangle = 0$$

which by applying the concept of reciprocity, yields

$$\left\langle\dfrac{\left\langle\dfrac{0}{1}\right\rangle}{\left\langle\dfrac{6}{1}\right\rangle}\right\rangle = 0.$$

By cross multiplying we obtain

$$\left\langle\dfrac{0}{1}\right\rangle = 0\left\langle\dfrac{6}{1}\right\rangle$$

which boils down to 0 = 0 and this is true and valid.

PROPERTIES OF DISTRIBUTION DENSITY (Operational Rules)

Let a, b, and c be non-zero, then

1.　$\left\langle\dfrac{a}{0}\right\rangle = -a$

2.　$-a\cdot\left\langle\dfrac{0}{b}\right\rangle = \dfrac{-a}{\left\langle\dfrac{b}{0}\right\rangle} = \left\langle\dfrac{-a}{\left\langle\dfrac{b}{0}\right\rangle}\right\rangle$. But $\left\langle\dfrac{b}{0}\right\rangle = -b$. Therefore,

　　$-a\cdot\left\langle\dfrac{0}{b}\right\rangle = \left\langle\dfrac{-a}{-b}\right\rangle = \dfrac{a}{b}$

　　where $a > 1$. This is the **Negative Zeroth-Exclusive Property**.

3.　$a\cdot\left\langle\dfrac{0}{b}\right\rangle = \dfrac{a}{\left\langle\dfrac{b}{0}\right\rangle} = \left\langle\dfrac{a}{\left\langle\dfrac{b}{0}\right\rangle}\right\rangle$. But $\left\langle\dfrac{b}{0}\right\rangle = -b$. Therefore, $a\cdot\left\langle\dfrac{0}{b}\right\rangle = \left\langle\dfrac{a}{-b}\right\rangle = \dfrac{a}{-b}$

　　where $a > 1$. This is the **Positive Zeroth-Exclusive Property**.

4.　Since $\left\langle\dfrac{a}{0}\right\rangle = -a$ and $\left\langle\dfrac{b}{0}\right\rangle = -b$, it implies that $a\left\langle\dfrac{b}{0}\right\rangle = \left\langle\dfrac{ab}{0}\right\rangle = -ab$.

　　This is the **Positive Zeroth-Inclusive Property**.

5.　$-a\left\langle\dfrac{b}{0}\right\rangle = \left\langle\dfrac{-ab}{0}\right\rangle = ab$. This is the **Negative Zeroth-Inclusive Property**.

6.　$\langle a\rangle = \left\langle\dfrac{a}{1}\right\rangle = a$.

7.　$\langle -a\rangle = -\left\langle\dfrac{a}{1}\right\rangle = -a$

9

8. $a \cdot \left\langle \dfrac{c}{b} \right\rangle = \dfrac{a}{\left\langle \dfrac{b}{c} \right\rangle} = \left\langle \dfrac{a}{\left\langle \dfrac{b}{c} \right\rangle} \right\rangle = \dfrac{a}{\dfrac{b}{c}} = \dfrac{ac}{b}$. This is the **Associative Property**.

9. $\left\langle \dfrac{a}{b} \right\rangle = \dfrac{\langle a \rangle}{\langle b \rangle} = \dfrac{a}{b}$.

10. $1 \cdot \left\langle \dfrac{a}{b} \right\rangle = \left\langle \dfrac{b}{a} \right\rangle^{-1} = \dfrac{1}{b/a} = \dfrac{a}{b}$. This is the **Unit Property**.

11. $0 \cdot \left\langle \dfrac{a}{b} \right\rangle = 0$. This is the **Null Property**.

12. $0 \cdot \left\langle \dfrac{a}{0} \right\rangle = 0$. This is the **Null Property**.

13. $0 \cdot \left\langle \dfrac{0}{b} \right\rangle = 0$. This is the **Null Property**.

14. $\left\langle \dfrac{0}{b} \right\rangle = 0$.

As shown in properties (2) and (3), the addition of an x number of $\left\langle \dfrac{0}{b} \right\rangle$ can be expressed as

$$x \cdot \left\langle \dfrac{0}{b} \right\rangle = \left\langle \dfrac{0}{b} \right\rangle_1 + \left\langle \dfrac{0}{b} \right\rangle_2 + \cdots + \left\langle \dfrac{0}{b} \right\rangle_x$$

where RHS in non-existent. In reality one cannot keep on adding zeros (i.e. $\left\langle \dfrac{0}{b} \right\rangle = 0$) to

gain incrementally. This case is called static summation (i.e. non-incremental addition). Dynamic summation (i.e. incremental addition) takes place only when an additional gain is experience. This is why reciprocity is used to attain a non-zero dividend in order to shift from a no-addition status to an addition status. From properties (4) and (5), the addition of x number of $\left\langle \dfrac{b}{0} \right\rangle$ which can be expressed as

$$x \cdot \left\langle \dfrac{b}{0} \right\rangle = \left\langle \dfrac{b}{0} \right\rangle_1 + \left\langle \dfrac{b}{0} \right\rangle_2 + \cdots + \left\langle \dfrac{b}{0} \right\rangle_x$$

where RHS represents an incremental addition scenario, is a case of dynamic summation.

MIXED FRACTION AND NUMERICAL ARITHMETIC REPRESENTATION

A numerical arithmetic representation can be written directly as a mixed fraction. For example, in case study B which is an absolute retention situation, we had

$$[6]\frac{6}{0} = 0.$$

The mixed fraction equivalent is given as $6\frac{0}{6} = 0$. Expressing the LHS as an improper fraction gives

$$\frac{0 \times 6 + 6}{0} = \frac{6}{0} = 0$$

This result can be made realistic by invoking property (1) to give

$$\left\langle \frac{6}{0} \right\rangle = -6.$$

Consequently, it could be said that the state of the distribution is equivalent to that of the recipients. That is, $-6 \equiv 0$ where -6 is the distributor's characteristic state relative to recipient(s) (if any) and 0 is the recipient(s) characteristic state relative to themselves. Also, in the no retention situation of case study A

$$[0]\frac{6}{3} = 2.$$

This can be expressed in mixed fractional form as

$$\frac{3 \times 0 + 6}{3} = \frac{6}{3} = 2$$

which is correct. Thus, $\left\langle \frac{6}{3} \right\rangle = 2$. For a last analysis here, let us consider a partial retention case where out of 7 items, 6 items are distributed among 3 students. The numerical quasi-arithmetic representation will be

$$[1]\frac{6}{3} = 2.$$

Expanding the LHS in terms of improper fraction gives

$$LHS = \frac{3 \times 1 + 6}{3} = \frac{9}{3} = 3.$$

Thus, $LHS \neq RHS$. Whereas one cannot say LHS is equal to the RHS, there exist a distribution balance (symbolized by $><$) here.

Generally,

> *If R_v, D_ρ, Q and r represent retention value, distributional density, quota (i.e. quotient) and the remainder respectively. Then,*
>
> *1. For cases of absolutely no retentions:* $\quad D_\rho = Q + r$
>
> *2. For cases of absolute retentions:* $\quad R_v + D_\rho >< Q + r$

THE CONCEPT OF DIVISION BY ZERO

Is a number divided by zero undefined, not possible or zero? To answer this fundamental question definitively there is the need to understand the term division in its entirety.

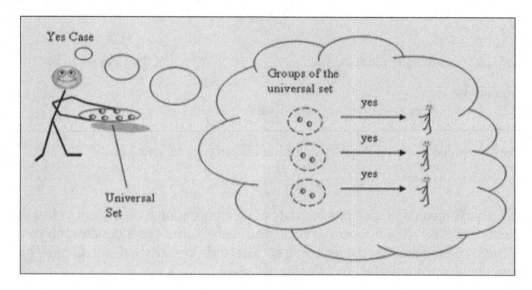

Figure 6. Equity distribution of six items.

An argument using sets or groups as a dividing process turns out to be very misleading. To verify this assertion, let us suppose one wants to divide (share) equitably 6 items among 3 would-be-recipients. One would have to group all 6 items into 3 and give each grouped item to one recipient. In the end, each of the 3 recipients would get 2 items as shown in figure 6. The assertion here is that,

Rule 1 (Ad hoc)

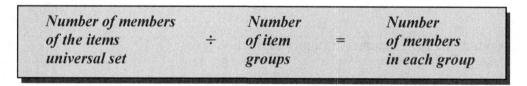

Number of members of the items universal set	÷	*Number of item groups*	=	*Number of members in each group*

where the RHS equal the amount each recipient gets. Upon this basis, the situation in figure 6 can be expressed as: $6 \div 3 = 2$ which it true. The value at the RHS of the equation (i.e. 2) is true in the prospective sense (i.e. amount received by each recipient). The next question relates to what happens if the 6 items where to be shared among nobody. Nobody here means no recipient gets anything even though they exist. The argument here is that if no recipient gets anything then there exists no grouping of items as shown in figure 7. Using the above set-flavoured division formula, the situation in figure 7 can be expressed as $6 \div 1 = 6$. This is sophistically true.

Specifically, it is true in the retrospective sense (that is the amount of items in the distributor's hand) if one consider the value at the RHS of the equation. Again, it must be

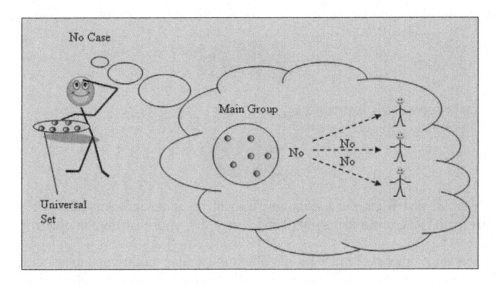

Figure 7. Selfish distribution of six items among no existing recipients.

borne in mind that in the scenario illustrated figure 7, each member of the recipient set has no item.

A second way of looking at the division assertion is to replace the divisor (i.e. the number of item groups) with the number of members of the recipient set who actually receive item(s) and that of the quotient as expressed in rule 2. Thus,

Rule 2 (Ad hoc)

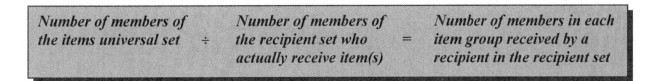

Number of members of the items universal set	÷	*Number of members of the recipient set who actually receive item(s)*	=	*Number of members in each item group received by a recipient in the recipient set*

Under the situation in figure 6, we get $6 \div 3 = 2$ and for that in figure 7 we get $6 \div 0 = 0$ which must be true in prospective sense. It is tempting to see the previous case of $6 \div 1 = 6$ as valid. However, remember that for the distributor to keep all 6 items solely because of unwillingness to share them with anybody cannot be considered as executing any transfer process of item(s) to another set of recipient(s). This is a case of non-division but an action refereed to as retention.

THE LAW OF EQUITY DISTRIBUTION

Under the condition of a whole number equity distribution, a novel mathematical operator for a sharing process called '**Equi-Distribution Operator'** will be introduced to buttress the results generally derived using the concept of distribution density.

In general, an ad hoc primary definition of the law of equity distribution is given by

$$\frac{X}{Y} = \langle Z \ \rho \ W \rangle$$

which can be secondarily expressed as

$$\frac{X}{Y} = Z + \frac{W}{Y}$$

where X, Y, Z, and W are the distributum, distribient, quota and remainder sets respectively, ρ is the remainder symbol and $< \rho >$ the 'equi-distribution operator'.

If the open angle bracket of the 'equi-distribution operator' is assigned to act as a multiplication operator on the quota term and the remainder symbol is also assigned to act as an addition operator on both the quota and remainder terms then,

$$\frac{X}{Y} = \langle Z \ \rho \ W \rangle$$

$$X = Y \langle Z \ \rho \ W \rangle$$

$$X = YZ + W$$

By dividing through both sides of the equation by Y we get

$$\frac{X}{Y} = \frac{YZ}{Y} + \frac{W}{Y} = Z + \frac{W}{Y}$$

Hence,

$$\frac{X}{Y} = \langle Z \ \rho \ W \rangle = Z + \frac{W}{Y}$$

which is the combined result sort.

A Quotient With Zero Numerator

An example of such a quotient will be: 0/2. Quantitatively, this means zero items are to be distributed between 2 people. This implies none of the 2 people get anything because there is nothing to share. As a result, there will be zero item left after nothing is there to share. By applying both primary and secondary versions of the ad hoc law of equity distribution and using the given sizes of corresponding sets involved, the above situation can be mathematically expressed jointly as

$$\frac{0}{2} = \langle 0 \ \rho \ 0 \rangle = 0 + \frac{0}{2} = 0$$

which is very true.

Division by Zero

The operation of dividing any number by zero distorts the uniqueness of division as an inverse operation of multiplication. Division by zero is therefore contemporary said to be undefined for real numbers. There are a few cases, however, where division by zero is considered as defined. The following are examples:

(i) $z/0$ for $z \in C^* \neq 0$ in the extended complex plane. Here, C-star (C^*) is called the complex infinity. Thus, the following fact is expressed:

$$\underset{w \to \infty}{Lim} \ \frac{z}{w} = 0 \quad \text{(complex infinity)}$$

The formal statement $1/0 = \infty$ is permitted in C-star, which implies $1 \neq 0 \cdot \infty$. Consequently, zero does not have a multiplicative inverse under any circumstances here. [2]

(ii)   Limits involving division by real number. For example, by definition

$$\underset{x \to 0}{Lim \ Sin} \frac{x}{x} = 1$$

At times, such limits may also approach infinity. For example

$$\underset{x \to 0^+}{Lim} \ \frac{1}{x} = \infty$$

As an illustrative example of division by zero, use will be made of 2/0. Using the primary ad hoc equation of the law of equity distribution, we get

$$\frac{2}{0} = \langle 0 \ \rho \ 2 \rangle = 0(0) + 2 = 2 .$$

This result is of course true! If two items are cut into zero parts, each part (which technically is the only existing parts) will be equal to the original items that never got divided and are still with the distributor. On the other hand, using the secondary ad hoc equation of the law of equity distribution gives

$$\frac{2}{0} = 0 + \frac{2}{0} = \frac{2}{0}$$

This result is tautological in nature. Previously, it was referred to as the '**paradoxical equivalence between division by zero and a zero division**' under '*The Zero Numerator and Zero Denominator Divisions Equivalence* '. Once again, a concise explanation and proof of the nothingness of dividends in both conditions (ii) and (iii) will be given under '*The Concept of Set Resonance*' and the '*Proof of Division by Zero Solution*' under '*The Logarithm of Division by Zero*'.

TWO BASIC NUMBER FIELDS

The tautology resulting from the use of the ad hoc equation of the law of equity distribution under division by zero can be eradicated. To do this let us, in terms of set sizes further analyze the secondary ad hoc equation of the law of equity distribution which is given as

$$\frac{x}{y} = z + \frac{w}{y}$$

where x, y, z, and w are the sizes of the distributum, distribient, quota and remainder sets respectively.

Notice from the above equation that

1. If y is equal to zero, the conventional situation of infinity arises. For example,

$$\frac{x}{0} = z + \frac{w}{0}$$

where w/0 represents division by zero.

2. If both w and y are each equal to zero, the following arises. For example,

$$\frac{x}{0} = z + \frac{0}{0}$$

where 0/0 represents another undefined fraction.

In condition (i), w/0 is temporary defined as a set of fused single numbers. However, in condition (ii), 0/0 presents a different case. This is due to the fact that the product of zero and any number gives zero. Thus, the solution for 0/0 is not unique. As such the value of z in condition (ii) will vary. Such serious mathematical aberration can be avoided by establishing the following formal definition for the '**Law of Equity Distribution**',

$$\frac{x}{y} = \langle z \ \rho \ w \rangle = z + \frac{w}{y}$$

$$provided \ \ y \neq w \ \ if \ \ y = 0.$$

Notice here that the specified condition attempts to prevent a situation where the remainder term, w could have been distributed among the recipient(s) or distribient without any fractional leftover but was not done. This is a case of illegitimate unwillingness to share on the part of the distributor.

Since there exists no distribient (y = 0), the remainder must be equal to the amount intended for distribution. Interestingly, y and w can never be the same else it will incidentally increase the whole number value of z.

To further analyze this current situation, let us create a series of outcomes from the formal definition for the 'Law of Equity Distribution' by solving for z using the ratio 2/0. With

$$x = 2$$

16

and

$$y = 0$$

we have

$$\frac{x}{y} = \frac{2}{0} = z + \frac{w}{0} \quad which \ gives \ z = \frac{2}{0} - \frac{w}{0}.$$

Since y = 0, w cannot be equal to y in accordance with the formal definition of the 'Law of Equity Distribution'. Subsequently, when

$$w = 1: \qquad z = \frac{2}{0} - \frac{1}{0} = \frac{1}{0}$$

$$w = 2: \qquad z = \frac{2}{0} - \frac{2}{0} = \frac{0}{0}$$

$$w = 3: \qquad z = \frac{2}{0} - \frac{3}{0} = \frac{-1}{0}$$

Notice here that the values of z were obtained by assuming that each division by zero term is a single number by itself. The set of abstract field axioms supporting this idea will be provided in due course.

Also, by multiply through the secondary ad hoc equation for the 'Law of Equity Distribution' we get

$$x = y\langle z \ \rho \ w \rangle = yz + w$$

which can be rearranged as

$$zy = x - w$$

Observe that always zy = 0 because y = 0 and from the equation for xy above, LHS is not equal to the RHS except when w = 2. Consequently, under the given conditions of x = 2, y = 2, and w = 1, ..., 3 the equation for zy can be correctly expressed as

$$zy \neq x - w$$

$$provided \ w \neq 2$$

This general lack of equality between the LHS and the RHS of the equation for zy prevents the determination of z by dividing the RHS by y. This is so because multiplication is not the reciprocal of division under the condition above. Consequently, the series of equations of z emanating from the various values of w exhibit the conditions of mathematical aberration due to the lack of unique solution (or the presence of multiple solutions) for each case. Thus, the entire ratio for which z is equal to should not be seen as a fraction but rather as a unit number dubbed a '**Non-Distributive Number**'. As an example, 1/0 means one entity not shared, 0/0 means zero entity not shared and −1/0 means a borrowed entity not shared.

17

Also given the ratio 0/2, the sequence of outcomes derived from the formal definition of the 'Law of Equity Distribution' by solving for z when x = 0 and y = 2 are given by

$$\frac{x}{y} = \frac{0}{2} = z + \frac{w}{2} \quad which \ gives \ z = \frac{0}{2} - \frac{w}{2}$$

Since y ≠ 0, w can be equal to y. Hence, when

$$w = 0: \qquad z = \frac{0}{2} - \frac{0}{2} = 0$$

$$w = 1: \qquad z = \frac{0}{2} - \frac{1}{2} = -\frac{1}{2}$$

$$w = 2: \qquad z = \frac{0}{2} - \frac{2}{2} = -1$$

$$w = 3: \qquad z = \frac{0}{2} - \frac{3}{2} = -\frac{3}{2}$$

Here 0, -1/2 , –1 and -3/2 are examples of '**Distributive Numbers**'. Whereas 0 represents a '**distributive zero**', 0/0 on the other hand represents a '**non-distributive zero**'. Generally, the distributive and non-distributive numbers are respectively represented symbolically as

$$\pm \ \Re \quad and \quad \dots \ \pm \ \Re$$

where **R** here represents the set of real numbers. In general, it can be stated that

A set of distributive numbers is said to constitute a real number field whereas a set of non-distributive numbers constitute an abstract number field.

Finally, for a realistic number system (which is essentially the set of real numbers, **R** or a real number field) to be generated, y from the given formal definition of the 'Law of Equity Distribution' should be a member of the set of integers, **I**. On the other hand, for an abstract number field, **A** to be generated, y should be zero.

THE ABSTRACT FIELD AXIOMS

By definition, the set of elements satisfying the field axioms for both addition and multiplication and is a commutative division algebra is referred to as a field.[3] Similarly, an abstract field can be created satisfying an abstract field axiom. These are defined below.

Name	Addition	Multiplication
Commutativity	$\dfrac{a}{0} + \dfrac{b}{0} = \dfrac{b}{0} + \dfrac{a}{0}$	$\dfrac{a}{0} \cdot \dfrac{b}{0} = \dfrac{b}{0} \cdot \dfrac{a}{0}$
Associativity	$\left(\dfrac{a}{0} + \dfrac{b}{0}\right) + \dfrac{c}{0} = \dfrac{a}{0} + \left(\dfrac{b}{0} + \dfrac{c}{0}\right)$	$\left(\dfrac{a}{0} \cdot \dfrac{b}{0}\right)\dfrac{c}{0} = \dfrac{a}{0}\left(\dfrac{b}{0} \cdot \dfrac{c}{0}\right)$
Distributivity	$\dfrac{a}{0}\left(\dfrac{b}{0} + \dfrac{c}{0}\right) = \dfrac{a}{0} \cdot \dfrac{b}{0} + \dfrac{a}{0} \cdot \dfrac{c}{0}$	$\left(\dfrac{a}{0} + \dfrac{b}{0}\right)\dfrac{c}{0} = \dfrac{a}{0} \cdot \dfrac{c}{0} + \dfrac{b}{0} \cdot \dfrac{c}{0}$
Identity	$\dfrac{a}{0} + \dfrac{0}{0} = \dfrac{a}{0} = \dfrac{0}{0} + \dfrac{a}{0}$	$\dfrac{a}{0} \cdot \dfrac{1}{0} = \dfrac{a}{0} = \dfrac{1}{0} \cdot \dfrac{a}{0}$
Inverses	$\dfrac{a}{0} + \left(\dfrac{-a}{0}\right) = \dfrac{0}{0} = \left(\dfrac{-a}{0}\right) + \dfrac{a}{0}$	$\dfrac{a}{0} \cdot \left(\dfrac{a}{0}\right)^{-1} = \dfrac{1}{0} = \left(\dfrac{a}{0}\right)^{-1} \cdot \dfrac{a}{0}$ $if \ \dfrac{a}{0} \neq \dfrac{0}{0}$

Table 1. The abstract field axioms.

These abstract field axioms would be applied where necessary in the topics that follow.

THE ZEROTH POWER OF ZERO

By contemporary definition, 0^0 is said to be undefined. However, such a rule is relaxed to simplify some formulas by defining it as:

1. $0^0 = 1$ [4]

2. $0^0 = 0$ [5]

The following proof is advanced to define 0^0.

Proof

By definition of indices,

$$x^{a-b} = \frac{x^a}{x^b}$$

Therefore,

$$x^0 = x^{1-1} = \frac{x^1}{x^1}$$

If x = 0, we get

$$0^0 = 0^{1-1} = \frac{0^1}{0^1}$$

The fair argument advocated here is that, *the evaluation of the respective numerator and denominator exponentiation must first be carried out followed by the division of the numerator by the denominator.* By definition, any number raised to the power 1 is equal to the same number. Hence

$$0^0 = \frac{0}{0}$$

This implies 0^0 is equal to an abstract zero.

THE QUESTION OF PREFERENCE

This is the willingness of the distributor to share or not share. Preference or choice can be seen from a distributor's or recipient's point of view. If the preference is one of a recipient's point of view, then it will be the recipient's willingness to accept an item(s).

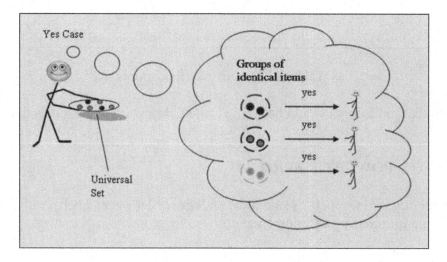

Figure 8. Multi-group illustration of specific equity distribution.

Using colour preferences by recipients as an illustrative example, the equity distribution of 6 items in figure 8 shows that each of the 3 recipients uniquely has a specific preference for red, blue or yellow but not a mixture of either colours.

Invoking rule 2, the situation in figure 8 can be expressed as $6 \div 3 = 2$. Now, what happens to the same 3 recipients if all of the 6 items are white in colour? While sharing action takes place by virtue of each recipient being offered 'items', the set of recipients notably have nothing of their preference to receive. So they receive nothing! In other words,

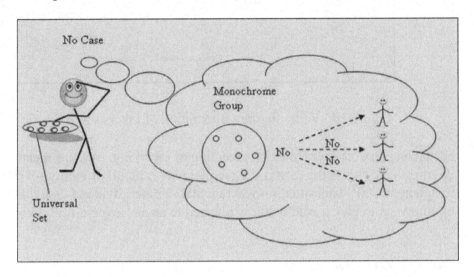

Figure 9. Mono-group illustration of specific equity distribution.

there is no recipient who actually receives any item in the end. Since 6/0 is a '**non-distributive number**', meaning six-offered items not shared, $6 \div 0 = \cdots 6$ (i.e. a positive non-distributive number) will be the numerical symbolism representative of the sharing process that takes place in the latter scenario. The illustration in figure 9 above depicts what happens. If it was the distributor who chose not to distribute the 6 items, then the situation would be represented as $6 \div 0 = \cdots - 6$. Notice that both numerical symbolic expressions give wholistic information about the sharing processes.

BASIC TYPES OF DIVISIONS

There exist six basic types of division that can occur in general distributive situations. These are:

A. *The Inverse-Null Division*

This occurs when no attempt is needed to be made at all to transfer items from the distributor to recipient(s). Using the case of 6 items, the null division can be illustrated in a Venn diagram as shown in figure 10. The situation is symbolically expressed as $6 \div \Phi = Void$. The term 'void' here means that no action of sharing

21

Figure 10. Venn diagram illustration of inverse-null division.

has taken place for there to exist a reaction of receiving. Also, φ means emptiness or empty set. This is the case if the recipient set is a null or empty set (i.e. no recipient exists). This situation where a real number dividend is divided by a null or empty set to give a void result is referred to as an 'inverse-null division'.

B. *The Non-Zero Division*

This type of division is the case where the distributor is willing to distribute all items. The illustrations in figures 6 and 8 are classic example. It is singularly represented by a Venn diagram in figure 11 below.

Figure 11. Venn diagram illustration of non-zero division.

C. *The Zilch Division (Division by Zero)*

This is a situation where a number of distributive items are distributed among zero number of recipients (i.e. nobody). The illustrations in figures 7 and 9 are classic example. It is singularly represented by a Venn diagram in figure 12 below.

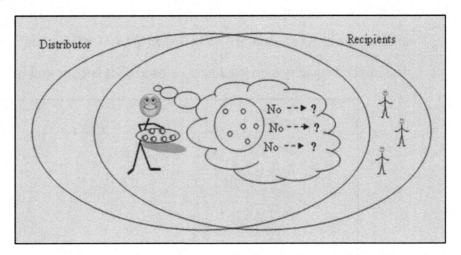

Figure 12. Venn diagram illustration of zilch division.

D. *The Zero-Dividend Division*

This is a situation where the number of distributive item to be distributed among a non-zero number of recipients is absolutely nothing (i.e. zero). The illustration in figure 5 is a classic example. Figure 13 below shows a Venn diagram

Figure 13. Venn diagram illustration of zero-dividend division.

representation of an example based on distributing a zero item among three recipients.

23

E. *The Redundant or Degenerated Division*

This is a situation where neither items nor recipients are available for distribution and receiving. Here the recipient(s) and distributor do not co-exist fully or partially. Figure 14 depicts a Venn diagram representation of the scenario. Symbolically, the scenario (based on rule 2) can be expressed as $0 \div 0 \Rightarrow= 0$ where $\Rightarrow=$ is an 'implied equality' sign. It is read, 'zero divided by zero equally implies zero'. As a result, if $\dfrac{0}{0} \Rightarrow= 0$ then $0 \Rightarrow= 0 \times 0 = 0$ which means $0 \Rightarrow= 0$.

The implied equality sign used here helps to alleviate the ambiguous nature of 0/0

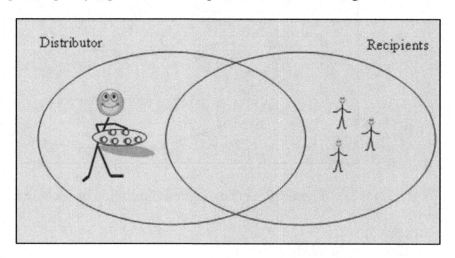

Figure 14. Venn diagram illustration of redundant division.

if an equal sign is used. Conventionally, with an equal sign it could be 'equal' to any number. Later on, this will be shown as wrong.

F. *The Null Division*

This is a reciprocated scenario of the inverse-null division. Here, no attempt needs to be made at all to receive items from a distributor. A Venn diagram representing this situation is shown in figure 15 below. A numerical symbolism for the scenario shown in figure 15 is $\Phi \div 3 = \Phi$. It should be noted that what the recipients received should be part of what is distributed (which is null in this situation).

It must be generally clear that an active sharing-interaction of distributable items always takes place when either distributor or recipient(s) or both exist in the intersection between distributor and recipient(s). In other words, the distributor and recipient(s)

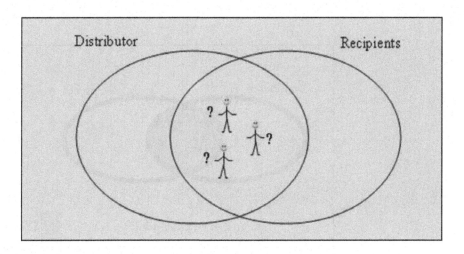

Figure 15. Venn diagram illustration of null division.

characteristically must co-exist completely or partially in a sharing-receiving process as shown in figure 16 below. The sharing-receiving region is known as

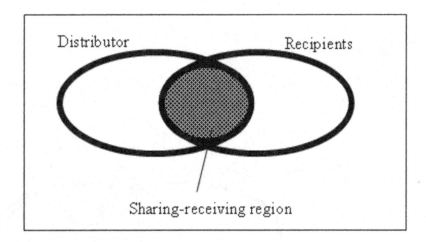

Figure 16.

the '**distributive region**' or '**distributive intersection**'.

POLARIZATION OF THE SHARING - RECEIVING PROCESS

The polarity of a division process is determined based upon the presence or absence of the distributor or recipient(s) or both in the sharing-receiving region. Invariably, there exist 3 main types of divisional polarities. These are listed below.

1. *Bi-Polar Division*

Here, both distributor and recipient(s) are found in the sharing-receiving region as

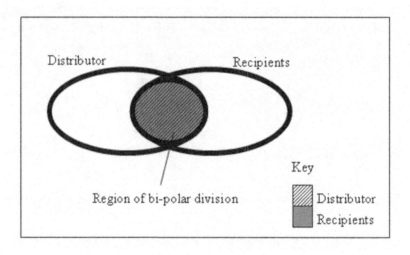

Figure 17. Illustration of bi-polar division.

depicted in figure 17 below.

2. *Mono-Polar Division*

Here, either distributor or recipient(s) is found in the sharing-receiving region. This situation in depicted in figures 18, 19, 20 and 21 below.

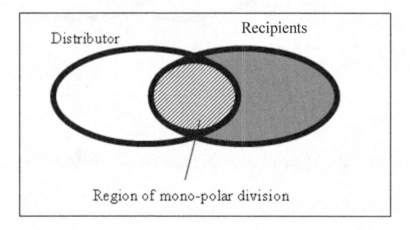

Figure 18. First illustration of mono-polar division.

The distributor and recipients are present here but only distributor exists in the

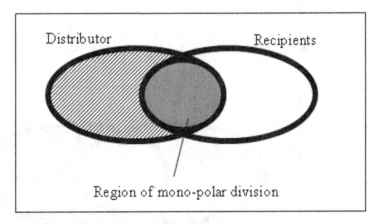

Figure 19. Second illustration of mono-polar division.

sharing-receiving region. The distributor and recipients are found here but only recipients exist in the sharing-receiving region.

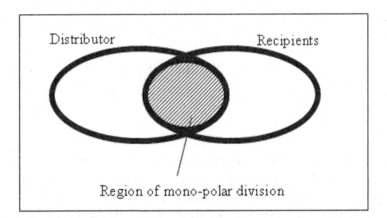

Figure 20. Third illustration of mono-polar division.

Only distributor is present here and exists in the sharing-receiving region.

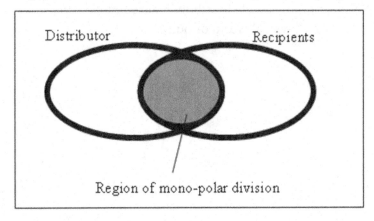

Figure 21. Fourth illustration of mono-polar division.

Only recipients are present here and exist in the sharing-receiving region.

27

3. *Non-Polar Division*

Here, neither the distributor nor the recipient(s) can be found in the sharing-

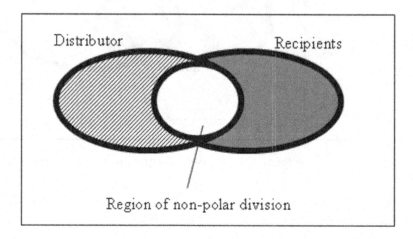

Figure 22. Illustration of non-polar division.

receiving region. The illustration for the non-polar condition is shown in figure 22 above.

The myriad nature of the types of divisions illustrated above using Venn diagrams, sensitizes how woefully inadequate the logics and symbolic language of mathematics has been applied to the description of the process of division. The lack of detail fundamental description(s) of the processes of division incubated and untimely gave rise to the numerous infamous mathematical undefinededness which riddles the world of mathematics. In chapter 5, '*Excogitating Unknown Mathematical Expressions*', proofs of solutions to such expressional situations will be formalized.

The section that follows will show how naturally these types of fundamental divisions permeate the very physical world of ours.

ATOMIC REMENISCENCE OF DIVISIONS

The set illustrations above depicting the types of divisions are reminiscent to the basic atomic or subatomic particles found in the world of matter and anti-matter.

The following convention is applied here. The distributor is hypothetically given a positive charge due to its sharing activity (i.e. giving) while the recipient(s) are ascribed a hypothetical negative charge. Table 2 below shows a comparative description of division based on the convention above.

FIGURE	ILLUSTRATIVE DESCRIPTION	HYPOTHETICAL ATOMIC/SUBATOMIC REMINISCENCE	COMMENT
10	Inverse-Null division	Proton	Has an outward electric field force line.
11	Non-Zero division	Neutron	Interaction of electric field force lines between electron(s) and proton(s) leads to permanent neutralization.
12	Division by zero	Atom	Interaction of electric field force lines between dynamic electron(s) and static proton(s) leads to partial neutralization.
13	Zero-dividend division	Anti-atom	Interaction of a reversed electric field force lines between dynamic positron(s) and static anti-proton(s) leads to a partial neutralization.
14	Redundant or Degenerated division	Atomic polarization / Polarized atom	Interaction of electric field force lines between excessive dynamic electron(s) and static proton(s) or vice versa leads to imbalanced neutralization.
15	Null division	Electron	Has an inward electric field force line.

Table 2. Types of division.

CHAPTER 2

GENERAL SURROGATION

SURROGATE OF THE RECIPEINTS' SET

This is an attempt to determine if there exists any surrogated receiving group(s) to especially take over the task of receiving when there exists no plausible member(s) in the set of recipients. It is obvious here that the members of the surrogated set must come from the region outside of the distributor and recipients sets. Thus, within an ideal universal set, U_i there exist the sets of distributor D, recipients R and surrogate recipients S. Note that the distributor set, D essentially is also the distributum set. On the other hand, a real universal set U_r will include only the sets of distributor and recipients.

Let the simplified probability ratio be P_o, x the number of distributive items in the distributive intersection, and y the number of recipients in the distributive intersection. Then

Condition A

$$\frac{x}{y} = \frac{n(d \in D \cap R)}{n(r \in D \cap R)} = P_o \cdot n(U_i) \tag{1}$$

where $P_o = P(D \cap R) = P(d \in D \cap R) \cdot P(r \in D \cap R)$. Also d and r are the distributable and recipients respectively. From the above equation (1), the amount of receivables per number of group(s) of recipients, R per G is given as

$$R \; per \; G = P_o \cdot n(r \in D \cap R) \cdot n(U_i) = n(d \in D \cap R) \tag{2}$$

Equations (1) and (2) are valid for conditions where

   i. All x and y exist in $D \cap R$.

   ii. Both x and y are not exiting in $D \cap R$.

By definition, P_o is found by dividing each numerator and denominator of the respective probability of d and r by the smaller of the dividend or divisor reacting in a distributive scenario.

Proof

By definition,

$$\frac{x}{y} = Q$$

where Q is the quotient, x the dividend and y the divisor. In terms of proportionality, $x \propto y$ which implies $x = Qy$ where Q is the constant of proportionality. Since x and y are also representative of the number of distributive items and number of recipients respectively, dividing each side of the proportionality expression by $n(U_i)$ gives

$$\frac{x}{n(U_i)} \propto \frac{y}{n(U_i)} \Rightarrow P'_x \propto P'_y$$

where P'_x and P'_y are the probability of x and y respectively. Thus, it could be said that

$$\frac{x}{n(U_i)} = Q\left(\frac{y}{n(U_i)}\right) = \left(\frac{Q}{n(U_i)}\right) \cdot y$$

Therefore,

$$x = \left(\frac{Q}{n(U_i)}\right) \cdot y \cdot n(U_i)$$

where $x = n(d \in D \cap R)$, $y = n(r \in D \cap R)$ and

$$\left(\frac{Q}{n(U_i)}\right) = P_o$$

QED (end of proof).

The above is the expression of equation (2). P_o is therefore called the '**quotient probability**'. It is found by simplifying the probability ratio for $(D \cap R)$. For example, let

$$P(D \cap R) = P(d \in D \cap R) \cdot P(r \in D \cap R) = \frac{6}{9} \cdot \frac{3}{9}$$

where $6 + 3 = n(U_r)$ or in general

$$n(U_r) = x + y.$$

Then,

$$P_o = \frac{x}{n(U_r)} \cdot \frac{y}{n(U_r)} = \frac{x' \cdot y'}{n(U_r)' \cdot n(U_r)'} = \frac{Q}{n(U_i)}$$

if and only if $\frac{x}{n(U_r)} = P_x$ or/and $\frac{y}{n(U_r)} = P_y$ exist (i.e. ≥ 0). In the equation for P_o, x, y

and $n(U_r)$ are divided by the smaller number among x and y to get $Q/n(U_i)$. If any of them is equal to 0, it implies it does not exist.

31

Condition B

For the case where n(D) or n(R) equals to 0

$$R \; per \; G = P_\in \cdot n(r \in D \cap R) \cdot n(U_i) = n(d \in D \cap R) \qquad (3)$$

where $P_\in$ is the existing quotient probability. If both or either of P_x or P_y do not exist, it is not accounted for (i.e. written or noted) at all.

By definition, the '**surrogate number**', Z is given by

$$Z = n(U_i) - n(D \cap R) \qquad (4)$$

where $n(U_i)$ is equal to the denominator of P_0. Z can also be called '**surrogate recipient(s)**' or '**surrogate quotient**'.

BASIC SECTIONS OF A DISTRIBUTIVE UNIVERSAL SET

The various parts of an ideal distributive universal set, U_i are illustrated in figure 23 below. Here x_a represents active distributive item(s), y_a represents active recipient(s) and $(D \cup R) \cup (D \cup R)'$ is the distributive region.

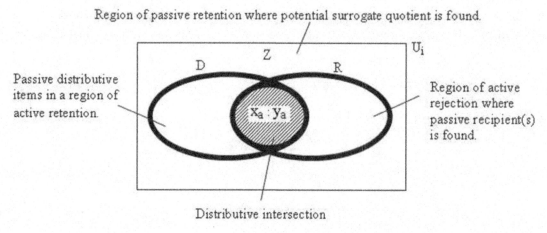

Figure 23. Elements of an ideal distributive universal set.

In the case of a real distributive universal set, U_r the various parts are shown in figure 24 below.

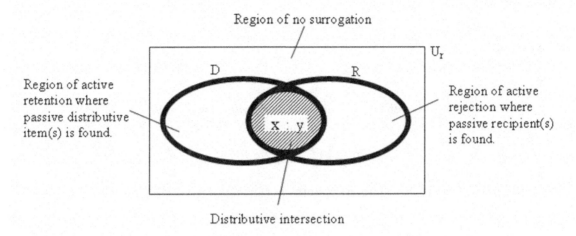

Region of no surrogation

Region of active
retention where
passive distributive
item(s) is found.

Region of active
rejection where
passive recipient(s)
is found.

Distributive intersection

Figure 24. Elements of a real distributive universal set.

Here, $D \cap R$ is the distributive region.

DETERMINATION OF SURROGATE NUMBERS

To better understand how to find the values of Z under ideal distributive universal set
condition, case studies will be used for purposes of analytical demonstration.

Case Study 1. How many of 6 active (or committed) items will each of 3 active recipients
get? The answer is 2 items each.

Analysis: The Venn diagram of the above problem is shown in figure 25. Here,
$n(D \cup R) = 9$. By definition,

$$P_o = P(D \cap R) = P(d \in D \cap R) \cdot P(r \in D \cap R)$$

$$= \frac{n(d \in D \cap R)}{n(D \cup R)} \cdot \frac{n(r \in D \cap R)}{n(D \cup R)}$$

$$= \frac{6}{9} \cdot \frac{3}{9} = \frac{6/3}{9/3} \cdot \frac{3/3}{9/3} = \frac{2 \cdot 1}{3 \cdot 3} = \frac{2}{9}.$$

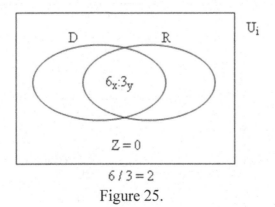

$6/3 = 2$

Figure 25.

33

This implies

$$n(U_i) = 9$$

under condition A.

Thus,

$$R \text{ per } G = P_o \cdot n(r \in D \cap R) \cdot n(U_i) = \frac{2}{9} \cdot 3 \cdot 9$$

which gives

$$R \text{ per } G = 2 \cdot 3.$$

The implication here is that there are 2 items per group of 3.

Also,

$$Z = n(U_i) - n(D \cap R) = 9 - 9 = 0$$

as shown in figure 25.

Case Study 2. How many of 6 active items will each non-participating recipients (i.e. zero recipient) get? The answer is 0.

Analysis: The Venn diagram of the above problem is shown in figure 26. Here,

$$n(D \cup R) = 9, \ n(d \in D \cap R) = 6, \ n(r \in D \cap R) = 0.$$

Under condition A,

$$P(D \cap R) = \frac{6}{9} \cdot \frac{0}{9} \ .$$

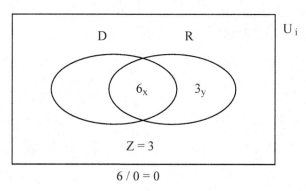

Figure 26.

Therefore,

$$P_o = \frac{6/3}{9/3} \cdot \frac{0/3}{9/3} = \frac{2 \cdot 0}{3 \cdot 3} = \frac{0}{9} \text{ and } n(U_i) = 9.$$

Hence,

$$R \text{ per } G = \frac{0}{9} \cdot 0 \cdot 9 = 0 \cdot 0.$$

This means 0 receivables per 0 group. Also, $Z = 9 - 6 = 3$.

Case Study 3. How many of 0 item (active) will each non-participating recipients get? The answer is 0.

Analysis: The Venn diagram of the above problem is shown in figure 27.

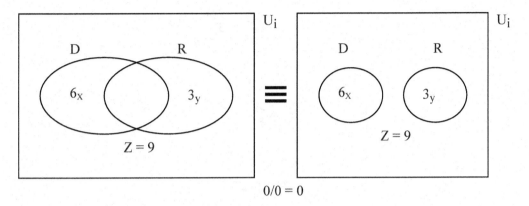

$$0/0 = 0$$

Figure 27.

Here,

$$n(D \cup R) = 9, \; n(d \in D \cap R) = 0, \; n(r \in D \cap R) = 0.$$

Under condition A,

$$P(D \cap R) = \frac{0}{9} \cdot \frac{0}{9}$$

Therefore,

$$P_o = \frac{0/3}{9/3} \cdot \frac{0/3}{9/3} = \frac{0 \cdot 0}{3 \cdot 3} = \frac{0}{9} \text{ and } n(U_i) = 9.$$

Hence,

$$R \text{ per } G = \frac{0}{9} \cdot 0 \cdot 9 = 0 \cdot 0$$

This means 0 receivables per group of 0. Also, Z = 9 – 0 = 9. Notice that D and R here are equivalent to disjointed sets. Now, let 0/0 = c where c is a real number. If c > 0 it will not be realistically true for c to be the quotient for 0/0 because no active recipient(s) exists. Therefore no items were received. Thus c must be 0. An explicit treatment of this case will be presented under '*Donor Surrogate and Unit Donor Distribution Principle*'. This is a case of an ideal socialist distribution (free-for-all). It occurs when there is a lack of total active distribution. The surrogation here is therefore an active one.

Case Study 4. How many of 4 active items will each of 2 active recipients get? The answer to this situation is 2.

Analysis: The Venn diagram of the above problem is shown in figure 28. Here,

$$n(D \cup R) = 6, \; n(d \in D \cap R) = 4, \; n(r \in D \cap R) = 2.$$

Under condition A,

$$P(D \cap R) = \frac{4}{6} \cdot \frac{2}{6} .$$

Therefore,

$$P_o = \frac{4/2}{6/2} \cdot \frac{2/2}{6/2} = \frac{2 \cdot 1}{3 \cdot 3} = \frac{2}{9}$$

and $n(U_i) = 9$.

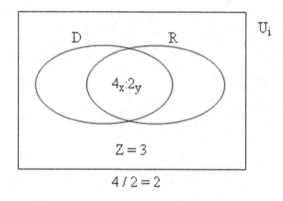

Figure 28.

Hence,

$$\text{R per G} = \frac{2}{9} \cdot 2 \cdot 9 = 2 \cdot 2 .$$

This means 2 receivables per group of 2. Also, $Z = 9 - 6 = 3$. The surrogation here is passive in nature.

Case Study 5. How many of 6 active items will each of 4 active recipients get? In this situation, the answer is 1½.

Analysis: The Venn diagram of the above problem is shown in figure 29. Here,

$$n(D \cup R) = 10, \ n(d \in D \cap R) = 6, \ n(r \in D \cap R) = 4 .$$

Under condition A,

$$P(D \cap R) = \frac{6}{10} \cdot \frac{4}{10} .$$

Therefore,

$$P_o = \frac{\frac{6}{4}}{\frac{10}{4}} \cdot \frac{\frac{4}{4}}{\frac{10}{4}} = \frac{\frac{3}{2} \cdot 1}{\frac{5}{2} \cdot \frac{5}{2}} = \frac{\frac{3}{2}}{6\frac{1}{4}}$$

and

$$n(U_i) = 6\frac{1}{4} .$$

36

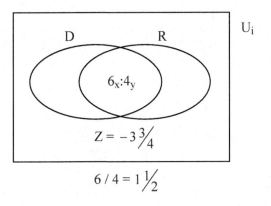

$$Z = -3\tfrac{3}{4}$$

$$6/4 = 1\tfrac{1}{2}$$

Figure 29.

Notice the importance of multiplying separately the denominators together. This is always the case. Hence,

$$\text{R per G} = \frac{\dfrac{3}{2}}{6\dfrac{1}{4}} \cdot 4 \cdot 6\dfrac{1}{4} = \frac{3}{2} \cdot 4 \,.$$

This means 3/2 receivables per group of 4. Also, $Z = 6\tfrac{1}{4} - 10 = -3\tfrac{3}{4}$.

Case Study 6. How many of 6 active items is each active member of the null or empty set going to get? The answer is -1.

Analysis: The Venn diagram of the above problem is shown in figure 30. Here,

$$n(D \cup R) = 6, \ n(d \ \ in \ \ D \cap R) = 6, \ n(r \ \ in \ \ D \cap R) = n(\Phi_a)$$

where $n(\Phi_a)$ is the active potential surrogate recipient number. Under condition B,

$$P(D \cap R) = \frac{6}{6} = P_\in \,.$$

So, if $Z = X$ we will have $n(U_i) = 6 + X$. Note that X is the potential surrogate recipient(s). Therefore,

$$\text{R per G} = \frac{6}{6} \cdot n(\Phi_a) \cdot (6 + X) = 6 \,.$$

This gives

$$n(\Phi_a) \cdot (6 + X) = 6$$

which results in

$$6 + X = \frac{6}{n(\Phi_a)} \,.$$

37

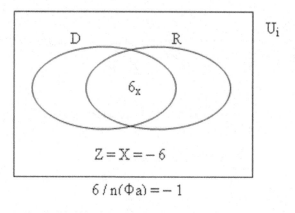

Figure 30.

But there exists no receivables at all since there is no active recipient(s). Thus,

$$6 + X = \frac{6}{n(\Phi_a)} \equiv \frac{6}{0} = 0.$$

This means $6 + X = 0$ and $X = -6$. This value of X represents the result for the distribution density

$$\left\langle \frac{6}{0} \right\rangle = \frac{\langle 6 \rangle}{\langle 0 \rangle} = -6$$

where $\langle 0 \rangle$ is equal to $n(\Phi_a)$ and -6 is the surrogate number or potential recipient(s). Later on, it will be shown that $n(\Phi_a) = Z = -6$ under '*The Characteristic States of the Surrogate Set*'.

CHARCTGERISTIC STATES OF THE SURROGATE SET

Generally, there are two possible states of surrogation. If Z is negative the surrogate recipient(s) are said to be active. On the other hand, a positive value of Z means the surrogate recipient(s) are passive.

Implications of Set R and Set S

Conventionally, the recipient null set of case study 6 is seen as non-existent and void. This neutral or passive null set concept is quantitative oriented. Let Φ be a non-existent null set. Then, if $R = \Phi$ it means that $n(\Phi) = 0$. This situation of non-existent null set is comparable to a void case of 'no school building(s) and no student(s) at all'. The lack of school building(s) represents the set R and the lack of student(s) is representative of the zero count of student population. The Venn diagram for the non-existent null set situation using case study 6 is shown in figure 31.

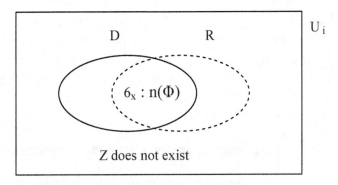

Figure 31. Case Study 6 revisited.

However, under the qualitative-quantitative oriented situation such as in case study 4, the **'null recipient set'**, Φ exists but not empty. This situation is reminiscent to the presence of a passive or active school building but with passive or active students respectively. If the null set of the recipient(s) is passive it means it has a quasi-existence and is represented as Φ_p.

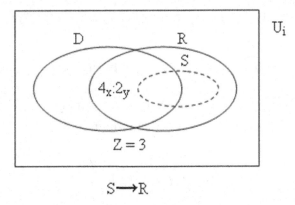

$$S \longrightarrow R$$

Figure 32. Case Study 4 revisited.

Here, the existence of the dormant surrogate set S singularly implies the existence of the set R as shown in figure 32. Details of the distribution of the members of the dormant set S

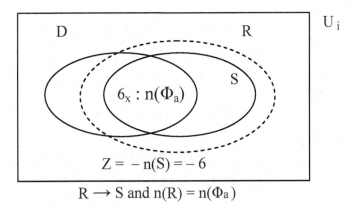

$$R \rightarrow S \text{ and } n(R) = n(\Phi_a)$$

Figure 33. Case Study 6 revisited.

will be shown later. On the other hand, if the null set of the recipient(s) is active within a given universal set, it does exist and it is represented by Φ_a. As already shown, the active null set Φ_a can be calculated for case study 6 which is represented by the Venn diagram shown in figure 33 below. Here, the set R is transformed to S, which exists actively. Thus R singularly implies S. The dichotomy between R and S is indicative of the possible existence of an active null set. In general,

> ***The set R implies the set S if and only if***
>
> ***1. The set R is a null set and***
>
> ***2. $n(S) < 0$ (i.e. active)***

It must be noted that the active recipient set and active null or surrogate set (i.e. $n(Z) < 0$) cannot both exist at the same time. Also, the passive or non-existent recipient set given by $n(R) = 0$ and the passive null or surrogate set also given by $n(Z) > 0$ cannot both coexist.

DISTRIBUTIVE PATTERN OF THE ACTIVE NULL'S SURROGATE SET

The members of the surrogate set S can be analyzed generally using the binomial theorem. Using equation (2) with the aid of the Venn diagram in figure 33, we get

$$\frac{6}{6 + n(\Phi_a)} \cdot \frac{n(\Phi_a)}{6 + n(\Phi_a)} \cdot n(\Phi_a) \cdot (6 + Z) = 6 .$$

This further gives

$$\frac{6n(\Phi_a)^2 (6 + Z)}{(6 + n(\Phi_a))^2} = 6$$

$$n(\Phi_a)^2 (6 + Z) = (6 + n(\Phi_a))^2$$

Thus

$$6n(\Phi_a)^2 + Zn(\Phi_a)^2 = 36 + 12n(\Phi_a) + n(\Phi_a)^2$$

$$(6 + Z - 1) \cdot n(\Phi_a)^2 - 12n(\Phi_a) - 36 = 0.$$

But $Z = -6$ (see figure 33). Therefore

$$- n(\Phi_a)^2 - 12n(\Phi_a) - 36 = 0$$

Using binomial theorem we get

$$n(\Phi_a) = \frac{-12 \pm \sqrt{144 - 4(36)}}{2} = \frac{-12 \pm 0}{2}$$

which gives

$$n(\Phi_a) = -6$$

This result is equal to the number of surrogate recipient(s), Z. Thus, it can in principle be stated that,

> **The number of surrogate recipient(s) is equal to the active null set's magnitude.**

This is mathematically expressed as

$$n(\Phi_a) = Z.$$

Consequently, the division taking place can be fully represented as

$$\frac{6}{n(\Phi_a)} = \frac{6}{-6} = -1$$

This result makes sense on equity basis. In a situation lacking recipient(s) completely and the willingness to equitably distribute 6 items, 6 surrogate recipients will be needed. This enables each surrogate recipient to receive one item. The negative sign of thee quotient above means it is surrogated and active as it was earlier on stated. Note that by extension, the active surrogate can be expressed as

$$Z' = n(U_i) - n(D \cup S)$$
$$= (6 + n(\Phi_a)) - (6 - 6)$$
$$= (6 - 6) - (6 - 6)$$
$$i.e. \ Z' = 0.$$

This means that the surrogate of recipients is not a chain reaction. Thus, in principle, it can be stated that

> **The surrogate set undergoes no further 'surrogation' or proxy processes.**

CHAPTER 3

SET PROBABILITY AND PROPENSITY

In contrast, the term probability connotes likelihood or possibility while propensity means predisposition, susceptibility or tendency. Thus propensity of an event occurring is a preconceived idea compared to its probability.

ROLE OF THE NULL SET

According to condition B, from the Venn diagram in figure 34, it is noted that

$$P(D \cap R) = P(D)$$

since R is an empty set. Therefore,

$$P(D \cap R) = \frac{n(D)}{n(D)} = \frac{6}{6} = 1.$$

Also,

$$n(r \ in \ D \cap R) = n(\Phi) \neq 0.$$

This is so because of the presence of empty recipient set, Φ which has neither active nor potential recipient(s)) within the empty recipient group. This situation is quite different from the existence of no active recipient(s) within a non-empty recipient group. It is also seen here that

$$n(U_i) = 6 + X$$

where X is the number of members of the compliment of D (i.e. D′). As it was shown earlier, using equation (2),

$$6 + X = \frac{6}{n(\Phi_a)} \equiv \frac{6}{0} = 0.$$

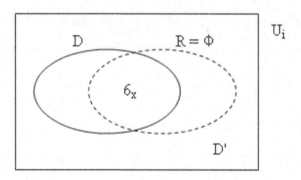

Figure 34. Case Study 6 revisited.

42

The LHS of the above expression can be written as

$$n(d \in D \cap R) + n(D') = \frac{n(d \in D \cap R)}{n(\Phi_a)}.$$

This implies that

$$n(U_i) = \frac{n(d \in D \cap R)}{n(\Phi_a)} \qquad (5)$$

Equation (5) stipulates that

> *The number of members of an active null set actively within a non-empty set is equal to the number of individual members in the entire ideal universal set under an equity distribution.*

What then does $n(\Phi_a)$ mean? To understand this, equation (5) is written as

$$n(\Phi_a) = \frac{n(d \in D \cap R)}{n(U_i)}$$

or it can be equivalently written (in terms of sets) as

$$\Phi \equiv \frac{D}{U_i}.$$

Since Φ is an empty set, it implies the ideal universal set does not exist in its subset. Also since the same is true for a real universal set, it can be stated that

$$U_i \equiv U_r.$$

Let U_n be a general universal set. Then it can be written that

$$n(\Phi_a) = \frac{n(d \in D \cap R)}{n(U_n)}.$$

But due to the presence of the null or empty set, by interacting with the distributor set, D it is found that

$$n(d \in D \cap R) = n(D).$$

Therefore,

$$n(\Phi_a) = \frac{n(D)}{n(U_n)} = P_n(D)$$

where P_n is the general probability of D. It is representative of the total probability existing in the universal set. The above result can be stated in terms of n(R) too (see Case Study 8).

43

Total Probability of an Ideal Situation

By definition, since n(D') is negative and the total members span $D \cup D'$, the probability due to this is given as

$$P(D) + P(D') = P(D) - P(D').$$

Thus

$$n_i(\Phi_a) = \frac{n(D)}{n(U_i)} - \frac{n(D')}{n(U_i)} = P_i(D_o)$$

where $n(D_o) = n(D) + n(D')$ $\;\;and\;\;$ $n(D_o)$ is the total members ideally existing in the ideal universal set. $P_i(D_o)$ is the ideal probability of $D \cup D'$.

It follows therefore that in principle

The surrogate probability is always negative. In other words, the surrogate state always has a negative probability.

In the case study under consideration (see figure 34), $n(D) = 6$ and $n(D') = |-6| = 6$. This is because members count is always represented by a positive number. Thus,

$$n(U_i) = 6 + 6 = 12.$$

Hence,

$$n(\Phi_a) = \frac{6}{12} - \frac{6}{12} = 0 = P_i(D_o)$$

Since,

$$P_i(D_o) = n_i(\Phi_a) = 0$$

it implies that Φ_a does not exist. As such it is shown in the Venn diagram of figure 35 as a broken circle.

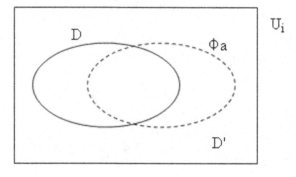

Figure 35.

Total Probability of a Real Situation

In real situations, the probability is mathematically expressed as,

$$n_r(\Phi_a) = \frac{n(D)}{n(U_r)} = P_r(D)$$

where $P_r(D)$ is the real probability of D. The above equation is so related because $n(D')$ here is equal to 0. Therefore

$$n(D_o) = n(D) + n(D') = n(D).$$

Using the same case study 6 (see figure 34) under consideration, $n(D) = 6$ and $n(U_r) = 6$. Therefore

$$n_r(\Phi_a) = \frac{6}{6} = 1.$$

This result implies Φ_a exists. The Venn diagram shown in figure 36 illustrates the situation.

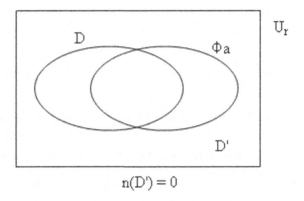

n(D') = 0

Figure 36.

Here, Φ_a is shown in continuous circle. Generally, n(Φ_a) represents the truth value of a set. It shows the degree of an existing receiving set becoming an empty set. This is true in nature. The idea of things/situations possibly fading away from perfection is realistic.

In general, it can be stated that

If R is not equal to a null set, then

1. $$n_i(\Phi_a) = \frac{n(D)}{n(U_i)} + \frac{n(D')}{n(U_i)}$$

2. $$n_r(\Phi_a) = \frac{n(D)}{n(U_r)}$$

PROPENSITY OF THE EXISTENCE OF THE RECIPEINT SET

The tendency of a distribution process becoming active is given by the '*Active Null Efficiency*' (**ANE**) measure. The ANE measures the propensity, $\widetilde{P}$ which is the natural tendency of the recipient set in a distribution to exist. By definition,

$$\widetilde{P} = ANE = \frac{n_r(\Phi_a)}{n_i(\Phi_a)} \times 100\%$$

By definition, while probability of an event taking place lies between 0 and 1 the propensity of that same event takes value from 0 and at best approaches 1 but can never be equal 1. The reason for this unique behaviour of propensity is lies in its formal definition (see above) where the real or actual number of Φ_a is always less than the ideal number of Φ_a.

Computation of Propensity for the Previous Case Studies

Case Study 1.

$$n_r(\Phi_a) = \frac{6}{9} = \frac{2}{3}; \quad n_i(\Phi_a) = \frac{6}{9} + \frac{3}{9} = 1. \quad \therefore \quad \widetilde{P} = \frac{n_r(\Phi_a)}{n_i(\Phi_a)} \times 100\% = \frac{2}{3} \times 100\% = 67\%.$$

The conclusion here is that the natural tendency for the distribution to continue in the same fashion is 67%. This means 2/3 of the active recipients will remain. However, with time 33% of the recipients will quit for example out of greed or for betterment. Here, Z is 0 because all items are distributed.

Case Study 2.

$$n_r(\Phi_a) = \frac{6}{9} = \frac{2}{3}; \quad n_i(\Phi_a) = \frac{6}{9} + \frac{3}{9} = 1. \quad \therefore \quad \widetilde{P} = \frac{n_r(\Phi_a)}{n_i(\Phi_a)} \times 100\% = \frac{2}{3} \times 100\% = 67\%.$$

There exist items to be distributed here. However, the existing set of recipient(s) dislikes the items. This state of distributive interaction will continue to oppose the items and 1/3 of recipients will quit opposing all together and participate in the distribution. Notice here that Z is 3 because there is room for a maximum number of 3 recipients.

Case Study 4.

$$n_r(\Phi_a) = \frac{4}{6} = \frac{2}{3}; \quad n_i(\Phi_a) = \frac{4}{9} + \frac{5}{9} = 1. \quad \therefore \quad \widetilde{P} = \frac{n_r(\Phi_a)}{n_i(\Phi_a)} \times 100\% = \frac{2}{3} \times 100\% = 67\%.$$

The conclusion here is the same as that of Case Study 1.

In the case studies above, notice that each would-be recipient receives 2 items. However, they all have the same propensity of 67%. It shows that in a dynamic situation of distribution (where both d and r exist in total or partial interaction), the recipient set will not exist always (i.e. $\widetilde{P} = 100\%$), but will tend to fade away in reality. This is true for scientific, business, political or sociological situations.

The natural tendency of a recipient remaining a recipient in a dynamic situation is referred to as '**apropensity**', $\overline{P}$. By definition

$$\overline{P} = 100\% - \widetilde{P}$$

However, in the case of R being equal to Φ (i.e. a null set), we have

$$\overline{P} = 100\% + \widetilde{P}$$

Case Study 5.

Here,

$$n_r(\Phi_a) = \frac{6}{10} = \frac{2}{5}; \quad n_i(\Phi_a) = \frac{6}{6\frac{1}{4}} + \frac{\frac{1}{4}}{6\frac{1}{4}} = 1.$$

$$\therefore \quad \widetilde{P} = \frac{n_r(\Phi_a)}{n_i(\Phi_a)} \times 100\% = \frac{3}{5} \times 100\% = 60\%.$$

The value of ANE is much smaller here. The implication is that lesser number of the active recipients will remain over the long haul. However, it will be later shown under '*The Distribution of Members of the Surrogate Set*' that the product of Z and $\widetilde{P}$ equals the quotient value which in this case is 1½.

Case Study 6.

As already computed (under 'Role of the Null Set'),

$$n_r(\Phi_a) = 1; \quad and \quad n_i(\Phi_a) = 0. \quad \therefore \quad \widetilde{P} = \frac{n_r(\Phi_a)}{n_i(\Phi_a)} \times 100\% = \frac{1}{0} \times 100\%.$$

$$But \quad \left\langle \frac{1}{0} \right\rangle = -1. \quad Hence \quad \widetilde{P} = -100\%.$$

This implies that $\widetilde{P} = 0\%$ and that the natural tendency of 'surrogation' is 100%. In other words, there is always a certainty for the surrogate population to become recipient(s).

The conclusion here is that the dynamic distribution is void under recipient group. However, it is possible under its surrogate distributive system.

Case Study 3.

This is a disjointed case. Finding propensity directly here is not possible. To circumvent this situation, use is made of other situations such as case studies 2 and 6. By definition

$$n_{iur}(\Phi_a) \text{ of case study } 3 = \left[1 - n_{iur}(\Phi_a) \text{ of case study } 6\right] \cap$$
$$\left[n_{iur}(\Phi_a) \text{ of case study } 2 - n_{iur}(\Phi_a) \text{ of case study } 6\right]$$

Note that the intersection is used because of events occurring at the same time. Hence, for the

Real case: $\qquad n_r(\Phi_a) = (1-1) \cap \left(\dfrac{2}{3} - 1\right) = 0 \cdot \left(-\dfrac{1}{3}\right) = 0$

Ideal case: $\qquad n_i(\Phi_a) = (1-0) \cap \left(\left(\dfrac{6}{9} + \dfrac{3}{9}\right) - 0\right) = 1 \times 1 = 1$

Therefore,

$$\widetilde{P} = \frac{0}{1} \times 100\% = 0\%.$$

This result means there is no null set formation here via the recipient set. The recipient set will always exist in a non-interactive manner. This shows how naturally unpleasant it is to see something you cannot get. The end result is to naturally quit!

The conclusion here is that because there is no distributive interaction, it is natural that with time all passive recipient(s) will quit for greener pastures. Z is 0 because no active division is taking place and so others are not prompted. In general, it can be stated that

$\widetilde{P}$ **takes values of Fibonacci fractions.**

Further Case Studies

Case Study 7.

How many of a zero item should each member of a null set get? The answer is zero.

Analysis: The Venn diagram of the above problem is shown in figure 37. Using equation (3) under condition B

$$P_x = \frac{6}{6} = 1, \ P_y = 0 \text{ and } n(U_i) = 6 + X \text{ where } X = n(D').$$

Notice that P_y is neglected. Therefore,

$$P_\in = P_x = 1.$$

Also,

$$n(r \in D \cap R) = n(\Phi) \equiv 0$$

because R is an empty set and $n(d \in D \cap R) = n(\Phi) \equiv 0$ because D is not an empty set.

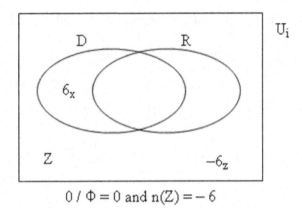

$0 / \Phi = 0$ and $n(Z) = -6$

Figure 37.

It must be observe that $n(\Phi)$ is the number of members in an empty set with no link to 'surrogation'. Therefore,

R per G equals $P_\in \cdot n(r \in D \cap R) \cdot n(\cup_i) = n(d \in D \cap R)$

which yields the following

$$1 \cdot n(\Phi) \cdot (6 + X) = 0$$

$$6 + X = \frac{0}{n(\Phi)} = 0.$$

Note that $n(\Phi) \equiv 0$. So, X = -6. That is, n(Z) = -6. To find the propensity,

$$n_r(\Phi_a) = \frac{6}{6} = 1$$

while

$$n_i(\Phi_a) = \frac{6}{0} - \frac{6}{0} = \left\langle \frac{6}{0} \right\rangle - \left\langle \frac{6}{0} \right\rangle = -6 - (-6) = 0.$$

Therefore,

$$\widetilde{P} = \frac{n_r(\Phi_a)}{n_i(\Phi_a)} \times 100\% = \frac{1}{0} \times 100\% = \frac{100}{0}\% = \left\langle \frac{100}{0} \right\rangle \% = -100\%.$$

Like the situation in Case Study 6, the set R is equal to the non-existent null set, Φ_0.

Case Study 8.

How many of a null set will each of 3 members of the set R get? In other words, 3 people wanting to receive but no ability to share exist. The answer is Φ.

Analysis: The Venn diagram of the above problem is shown in figure 38. Using equation (3) under condition B: $P_x = \dfrac{0}{3} = 0,\ P_y = \dfrac{3}{3} = 1 = P_\epsilon,\ n(U_i) = 3 + X,\ n(r \in D \cap R) = 3$ and $n(d \in D \cap R) = n(\Phi) \equiv 0$. Thus, R per G equals

$$P_\epsilon \cdot n(r \in D \cap R) \cdot n(U_i) = n(d \in D \cap R)$$
$$1 \cdot 3 \cdot (3 + X) = n(\Phi)$$
$$3 + X = \frac{n(\Phi)}{3}$$
$$3 + X = \frac{0}{3}$$
$$X = -3.$$

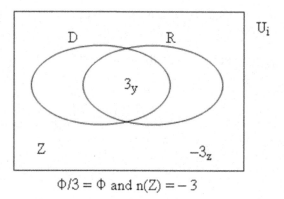

$\Phi/3 = \Phi$ and $n(Z) = -3$

Figure 38.

This implies $n(Z) = -3$. Here, $n(Z)$ represents surrogate member(s) for the set D and not R. To find the propensity, since D here has no members,

$$n_r(\Phi_a) = \frac{n(R)}{n(U_r)}$$

where

$$n(U_r) = n(D \cup R) \text{ and } n_i(\Phi_a) = \frac{n(R)}{n(U_i)} - \frac{n(R')}{n(U_i)}.$$

Therefore,

$$n(U_i) = n(D \cup R) + n(D \cup R)' = 3 - 3 = 0$$

50

and $n(U_r) = n(D \cup R) = 3$, $n(R) = 3$ and $n(R)' = -3$. Hence, $n_r(\Phi_a) = \dfrac{3}{3} = 1$ and

$$n_i(\Phi_a) = \frac{3}{0} - \frac{3}{0} = \left\langle \frac{3}{0} \right\rangle - \left\langle \frac{3}{0} \right\rangle = -3 - (-3) = 0.$$

The propensity is therefore given as

$$\widetilde{P} = \frac{n_r(\Phi_a)}{n_i(\Phi_a)} \times 100\% = \frac{1}{0} \times 100\% = \frac{100}{0}\% = \left\langle \frac{100}{0} \right\rangle \% = -100\%.$$

It means, even though the set R has 3 active members it is already a non-existent null set, Φ. Consequently, having recipients with nothing to gain is like they being non-existent.

Case Study 9: The meaning of negative zero.

How many of one null set will 'each' of zero recipient get? The answer is a null set, Φ.

Analysis: The Venn diagram of the above problem is shown in figure 39. Thus

$$P_x = \frac{0}{3} = 0, \; P_y = \frac{3}{3} = 1 = P_\in \, , \; n(U_i) = 3 + X \, , \; n(r \in D \cap R) = 0 \text{ and } n(d \in D \cap R) = n(\Phi).$$

Using equation (3) under condition B, R per G equals

$$P_\in \cdot n(r \in D \cap R) \cdot n(U_i) = n(d \in D \cap R)$$
$$1 \cdot 3 \cdot (3 + X) = n(\Phi).$$

This implies $n(\Phi) = 0$ which is true. On the other hand, if the equation for R per G is written as

$$P_\in \cdot n(U_i) = \frac{n(d \in D \cap R)}{n(r \in D \cap R)}$$

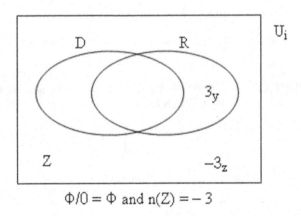

$$\Phi / 0 = \Phi \text{ and } n(Z) = -3$$

Figure 39

51

The negative sign here means that the set D (which is a null set) is susceptible to 'surrogation'. Hence, $3 + X = 0$ and $X = -3$. This implies n(Z) which is equal to -3 is the size of the surrogate members for the set D and not set R. To find the propensity, we use

$$n_r(\Phi_a) = \frac{n(R)}{n(U_r)} = \frac{3}{3} = 1$$

and

$$n_i(\Phi_a) = \frac{n(R)}{n(U_i)} - \frac{n(R')}{n(U_i)} = \frac{3}{0} - \frac{3}{0}$$

$$= \left\langle \frac{3}{0} \right\rangle - \left\langle \frac{3}{0} \right\rangle$$

$$= -3 - (-3)$$

$$= 0.$$

Therefore,

$$\widetilde{P} = \frac{n_r(\Phi_a)}{n_i(\Phi_a)} \times 100\% = \frac{1}{0} \times 100\% = \frac{100}{0}\% = \left\langle \frac{100}{0} \right\rangle\% = -100\%.$$

This implies that the set R is already a non-existent null set, Φ even though it has 3 passive members.

DISTRIBUTION OF MEMBERS OF THE SURROGATE SET

Knowing the portion of the members of the surrogate set that are active and passive within the 'surrogation' is very paramount. This helps one to envisage the surrogate process in the light of seemingly insurmountable distributive process(es) such as division by zero.

There two important products here that one needs to understand. These are

1. $Z\widetilde{P}$. This is a measure of the number of active surrogate recipient(s).

2. $Z\overline{P}$. This is a measure of the number of passive surrogate recipient(s).

The table 3 below shows the nature of typical distributions.

	ACTIVE DISTRIBUTION X:Y	NET SURROGATE RECIPIENT(S) Z	NUMBER OF ACTIVE SURROGATE RECIPIENT(S) $Z\widetilde{P}$	NUMBER OF PASSIVE SURROGATE RECIPEINT(S) $Z\overline{P}$
A	6:3	0	$0 \cdot \frac{2}{3} = 0$	$0 \cdot \frac{1}{3} = 0$
B	4:2	3	$3 \cdot \frac{2}{3} = 2$	$3 \cdot \frac{1}{3} = 1$
C	6:0	-6	$-6 \cdot \frac{1}{0} = -\frac{-6}{0} = \left\langle \frac{-6}{0} \right\rangle = 6$	$-6 \cdot \frac{0}{1} = 0$
D	6:4	$-3\frac{3}{4}$	$-3\frac{3}{4} \cdot \frac{3}{5} = \left\langle \frac{-9}{4} \right\rangle = \frac{9}{4}$	$-3\frac{3}{4} \cdot \frac{2}{5} = \left\langle \frac{-3}{2} \right\rangle = \frac{9}{4}$

Table 3.

Observe that $X \in D$, $Y \in R$ and $Z \in S$. Also,

$$Z\tilde{P} + Z\overline{P} = Z.$$

The Venn diagrams for the respective distributions (i.e. A, B and C) are shown below in figures 40 to 42 respectively.

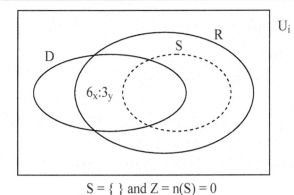

$$S = \{ \ \} \text{ and } Z = n(S) = 0$$

Figure 40. Distribution A.

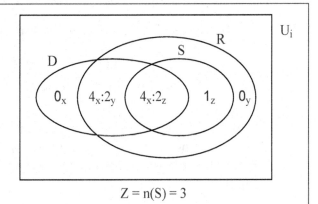

$$Z = n(S) = 3$$

Figure 41. Distribution B.

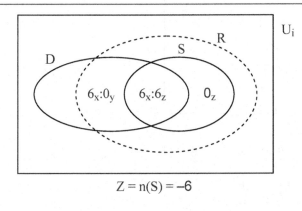

$$Z = n(S) = -6$$

Figure 42 a. Distribution C.

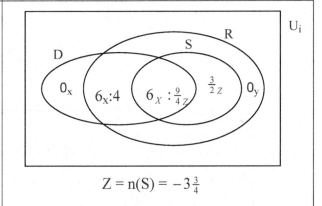

$$Z = n(S) = -3\tfrac{3}{4}$$

Figure 42 b. Distribution D.

In general, the following can be said about surrogate distribution.

If under a surrogate distribution,

$$n(Z) \equiv n(U_i)$$

then the maximum value between $Z\tilde{P}$ and $Z\overline{P}$ represents the quotient value of the distributive process.

53

Meaning of Z

Z means the number of surrogate recipient(s) that will evolve as the recipient(s) set R diminishes to a null set.

There are three possible kinds of Zs. If Z is

1. Negative, it implies active surrogate recipient(s).
2. Positive, it implies passive surrogate recipient(s).
3. Neutral (i.e. zero), it implies a non-existing surrogate recipient(s).

Let, X_p be the number of passive distributive items, X_a the number of active distributive items, Y_a the number of active recipient(s), Y_p the number of passive recipient(s), Z_a the number of active surrogate recipient(s) and Z_p the number of passive surrogate recipient(s). Then, the

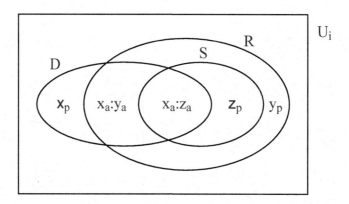

Figure 43. Venn diagram illustration of a surrogate set.

distribution of members of the surrogate set can be generally illustrated in a Venn diagram as shown in figure 43 above.

A Universal Redistribution Analysis

Under '*Determination of Surrogate Numbers*', case study 3 offers an ideal case to fundamentally analyze the benefit of redistributing items to all members of the universal set. As the case was, with the lack of total active distribution, the surrogate members, $n(Z) = 9$ will serve as the ideal recipients. Thus, $n(D) = 6$ items will be actively distributed to 9 surrogate members representing the size of the universal set. Hence, the distributive process (i.e. sharing) is expressed as 6/9 which is equal to 2/3. The result is less than what would have been the quotient if the receiving members of the receiving set, R engaged in active distribution which is represented as 6/3 = 2.

By definition, the propensity under universal redistribution, is given by

$$\tilde{P} = \frac{n_r(\Phi_a)}{n_i(\Phi_a)} \times 100\%.$$

But,

$$n_r(\Phi) = \frac{6}{15} \quad and \quad n_i(\Phi) = \frac{6}{15} + \frac{9}{15} = 1.$$

Therefore,

$$\tilde{P} = \frac{6}{15} \times 100\% = 40\%.$$

On the other hand, the would-be propensity for the case of distribution between recipients computes to

$$\tilde{P} = \frac{6}{9} \times 100\% = 67\%.$$

Evidently, it can be stated here that

> A universal redistribution fundamentally lacks adequate natural tendency to take place.

This explains why the socialistic idea of redistribution always lacks natural persuasion but driven by an engine of undesired force. Consequently, it is naturally bound to fail.

NATURE OF CONDITIONAL POWER SET OF THE RECIPIENT SET

The general definition for a conditional power set of the recipient set can be state as follows.

> *If R and S exist and are*
>
>   *1. complement sets (i.e. $S \subset R$ as they always are),*
>
>   *2. $R \Rightarrow S$ and*
>
>   *3. R is a non-existence set*
>
> *then R is said to be a power set.*

The reason for the above statement is that the members of the set R (as stated above) are always together as a unit in the active region of a distribution. This is the situation in case study 6. However, there exists a situation, such as case study 4, where the above statement is not valid.

THEOREMS OF MULTIPLE SURROGATES AND MULTIPLE PROPENSITIES

In arithmetic, under division by grouping it is established that,

> *If two numbers are divisible by the same number,*
> *then their sum is also divisible by that number.*

This is the theorem of grouping the dividend. For example,

$$15/3 = 5 \text{ and } 21/3 = 7.$$

Therefore

$$36/3 = 5 + 7 = 12.$$

The theorem of grouping the dividend will be analyzed on the basis of distribution sets. For each active distribution x/y analyzed below, the corresponding quotient probability P_o, surrogate number Z, real active surrogate number $n_r(\Phi_a)$, ideal active surrogate number $n_i(\Phi_a)$ and propensity $\widetilde{P}$ are computed.

A. For $\dfrac{6}{2}$:
$$P_o = \frac{Q}{n(U_i)} = \frac{n(D)}{n(U_r)} \cdot \frac{n(R)}{n(U_r)} = \frac{6}{9} \cdot \frac{3}{9} = \frac{2}{3} \cdot \frac{1}{3} = \frac{2}{9}$$

where $n(U_i) = 9$ from the denominator of P_o and $n(U_r) = x + y = 9$.

Also,
$$Z = n(U_i) - n(U_r) = 9 - 9 = 0.$$

This implies
$$n_r(\Phi_a) = \frac{n(D)}{n(U_r)} = \frac{6}{9} = \frac{2}{3}$$

and
$$n_i(\Phi_a) = \frac{n(D)}{n(U_i)} + \frac{n(D')}{n(U_i)} = \frac{6}{9} + \frac{3}{9} = 1$$

where $n(D') = n(U_r) - n(D)$.

Therefore
$$\widetilde{P} = \frac{n_r(\Phi_a)}{n_i(\Phi_a)} = \frac{2}{3}.$$

56

Figure A below depicts the scenario.

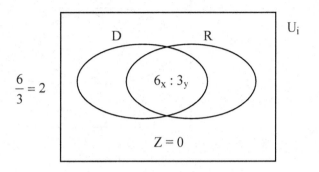

Figure A

B.   For $\dfrac{9}{3}$: $P_o = \dfrac{Q}{n(U_i)} = \dfrac{n(D)}{n(U_r)} \cdot \dfrac{n(R)}{n(U_r)} = \dfrac{9}{12} \cdot \dfrac{3}{12} = \dfrac{3}{4} \cdot \dfrac{1}{4} = \dfrac{3}{16}$

where $n(U_i) = 16$ and from the denominator of P_o $n(U_r) = x + y = 12$.

Also

$$Z = n(U_i) - n(U_r) = 16 - 12 = 4.$$

This implies

$$n_r(\Phi_a) = \dfrac{n(D)}{n(U_r)} = \dfrac{9}{12} = \dfrac{3}{4}$$

and

$$n_i(\Phi_a) = \dfrac{n(D)}{n(U_i)} + \dfrac{n(D')}{n(U_i)} = \dfrac{9}{16} + \dfrac{7}{16} = 1$$

where $n(D') = n(U_i) - n(D)$.

Therefore

$$\widetilde{P} = \dfrac{n_r(\Phi_a)}{n_i(\Phi_a)} = \dfrac{3}{4}.$$

Figure B below depicts the scenario.

57

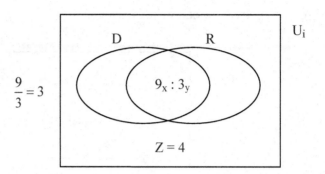

Figure B

C. For $\dfrac{12}{3}$: $P_o = \dfrac{Q}{n(U_i)} = \dfrac{n(D)}{n(U_r)} \cdot \dfrac{n(R)}{n(U_r)} = \dfrac{12}{15} \cdot \dfrac{3}{15} = \dfrac{4}{5} \cdot \dfrac{1}{5} = \dfrac{4}{25}$

where $n(U_i) = 16$ and from the denominator of P_o, $n(U_r) = x + y = 12$.

Also

$$Z = n(U_i) - n(U_r) = 25 - 15 = 10.$$

This implies

$$n_r(\Phi_a) = \dfrac{n(D)}{n(U_r)} = \dfrac{12}{15} = \dfrac{4}{5}$$

and

$$n_i(\Phi_a) = \dfrac{n(D)}{n(U_i)} + \dfrac{n(D')}{n(U_i)} = \dfrac{12}{25} + \dfrac{13}{25} = 1$$

where $n(D') = n(U_i) - n(D)$.

Therefore,

$$\widetilde{P} = \dfrac{n_r(\Phi_a)}{n_i(\Phi_a)} = \dfrac{4}{5} \div 1 = \dfrac{4}{5}.$$

Figure C below depicts the scenario.

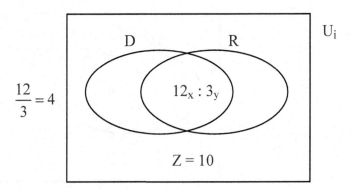

Figure C

D. For $\dfrac{4}{2}$: $P_o = \dfrac{Q}{n(U_i)} = \dfrac{n(D)}{n(U_r)} \cdot \dfrac{n(R)}{n(U_r)} = \dfrac{4}{6} \cdot \dfrac{2}{6} = \dfrac{2}{3} \cdot \dfrac{1}{3} = \dfrac{2}{9}$

where $n(U_i) = 9$ and from the denominator of P_o $n(U_r) = x + y = 6$.

Also
$$Z = n(U_i) - n(U_r) = 9 - 6 = 3.$$

This implies
$$n_r(\Phi_a) = \dfrac{n(D)}{n(U_r)} = \dfrac{4}{6} = \dfrac{2}{3}$$

and
$$n_i(\Phi_a) = \dfrac{n(D)}{n(U_i)} + \dfrac{n(D')}{n(U_i)} = \dfrac{4}{9} + \dfrac{5}{9} = 1$$

where $n(D') = n(U_i) - n(D)$.

Therefore
$$\widetilde{P} = \dfrac{n_r(\Phi_a)}{n_i(\Phi_a)} = \dfrac{2}{3}.$$

Figure D below depicts the scenario.

59

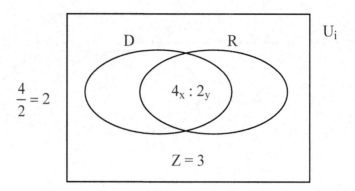

$$\frac{4}{2} = 2$$

Figure D

E.   For $\dfrac{6}{2}$: $P_O = \dfrac{Q}{n(U_i)} = \dfrac{n(D)}{n(U_r)} \cdot \dfrac{n(R)}{n(U_r)} = \dfrac{6}{8} \cdot \dfrac{2}{8} = \dfrac{3}{4} \cdot \dfrac{1}{4} = \dfrac{3}{16}$

where $n(U_i) = 16$ and from the denominator of P_o $n(U_r) = x + y = 8$.

Also

$$Z = n(U_i) - n(U_r) = 16 - 8 = 8 .$$

This implies

$$n_r(\Phi_a) = \frac{n(D)}{n(U_r)} = \frac{6}{8} = \frac{3}{4}$$

and

$$n_i(\Phi_a) = \frac{n(D)}{n(U_i)} + \frac{n(D')}{n(U_i)} = \frac{6}{16} + \frac{10}{16} = 1$$

where $n(D') = n(U_i) - n(D)$.

Therefore

$$\widetilde{P} = \frac{n_r(\Phi_a)}{n_i(\Phi_a)} = \frac{3}{4} .$$

Figure E below depicts the scenario.

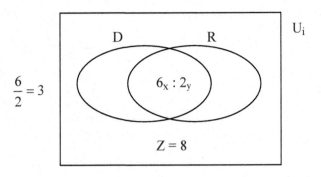

$\dfrac{6}{2} = 3$ (D ⋯ R) $6_x : 2_y$ U_i

$Z = 8$

Figure E

F.   For $\dfrac{15}{3} = \dfrac{6}{3} + \dfrac{9}{3}$: $P_0 = \dfrac{Q}{n(U_i)} = \dfrac{n(D)}{n(U_r)} \cdot \dfrac{n(R)}{n(U_r)} = \dfrac{15}{18} \cdot \dfrac{3}{18} = \dfrac{5}{6} \cdot \dfrac{1}{6} = \dfrac{5}{36}$

where $n(U_i) = 36$ and from the denominator of P_0 $n(U_r) = x + y = 18$.

Also

$$Z = n(U_i) - n(U_r) = 36 - 18 = 18 .$$

This implies

$$n_r(\Phi_a) = \dfrac{n(D)}{n(U_r)} = \dfrac{15}{18} = \dfrac{5}{6}$$

and

$$n_i(\Phi_a) = \dfrac{n(D)}{n(U_i)} + \dfrac{n(D')}{n(U_i)} = \dfrac{15}{36} + \dfrac{21}{36} = 1$$

where $n(D') = n(U_i) - n(D)$.

Therefore

$$\widetilde{P} = \dfrac{n_r(\Phi a)}{n_i(\Phi a)} = \dfrac{5}{6}$$

Figure F below depicts the above scenario.

61

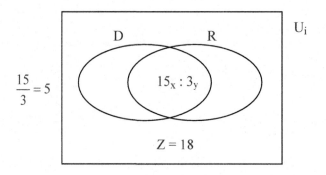

$$\frac{15}{3} = 5$$

Z = 18

Figure F

The table below includes results from computations done above.

Example	Active Distribution $\frac{x}{y}$	Quotient Probability P_o	Surrogate Number Z	Real Active Surrogate Number $n_r(\Phi_a)$	Ideal Active Surrogate Number $n_i(\Phi_a)$	Propensity $\widetilde{P}$
A	$\frac{6}{3}$	$\frac{2}{9}$	0	$\frac{2}{3}$	1	$\frac{2}{3}$
B	$\frac{9}{3}$	$\frac{3}{16}$	4	$\frac{3}{4}$	1	$\frac{3}{4}$
C	$\frac{12}{3}$	$\frac{4}{25}$	10	$\frac{4}{5}$	1	$\frac{4}{5}$
D	$\frac{4}{2}$	$\frac{2}{9}$	3	$\frac{2}{3}$	1	$\frac{2}{3}$
E	$\frac{6}{2}$	$\frac{3}{16}$	8	$\frac{3}{4}$	1	$\frac{3}{4}$
F	$\frac{15}{3}$	$\frac{5}{36}$	18	$\frac{5}{6}$	1	$\frac{5}{6}$
G	$\frac{10}{2}$	$\frac{5}{36}$	24	$\frac{5}{6}$ *	1 *	$\frac{5}{6}$ *
H	$\frac{21}{3}$	$\frac{7}{64}$	40	$\frac{7}{8}$ *	1 *	$\frac{7}{8}$ *
* See Venn diagrams below for computations.						

Table 4.

From the active distribution A, B and F it is seen that, if their respective surrogate numbers are Z_1, Z_2 and Z_3, then $Z_3 \neq Z_1 \cdot Z_2$ *i.e.* $18 \neq 0 \cdot 4$. However, from D, E and G if their respective surrogate numbers are Z_1, Z_2 and Z_3, then $Z_3 = Z_1 \cdot Z_2$ *i.e.* $24 = 3 \cdot 8$. Also for

B, C and H, if their respective surrogate numbers are Z_1, Z_2 and Z_3, then $Z_3 = Z_1 \cdot Z_2$ i.e. $40 = 4 \cdot 10$.

Consequently, it can be generally be stated that

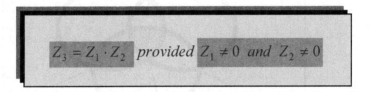

$$Z_3 = Z_1 \cdot Z_2 \quad provided \quad Z_1 \neq 0 \quad and \quad Z_2 \neq 0$$

The above result is not surprising since the theorem of grouping the dividend could be restated as,

If two numbers are multiples of the same number,
Then their sum is also a multiple of that number.

Following the quantitative analysis of the theorem of grouping the dividend, the situations in D, E and G can be generally depicted as shown below in figure 44.

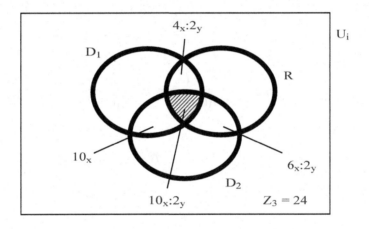

Figure 44.

For G:

$$n_r(\Phi_a) = \frac{n(D_1 \cup D_2)}{n(U_r)} = \frac{10}{12} = \frac{5}{6}$$

and

$$n_i(\Phi_a) = \frac{n(D_1 \cup D_2)}{n(U_i)} + \frac{n(D_1 \cup D_2)'}{n(U_i)} = \frac{10}{36} + \frac{26}{36} = 1.$$

Therefore

$$\widetilde{P} = \frac{5}{6}.$$

63

On the other hand, the situations in B, C and H are generally depicted in figure 45.

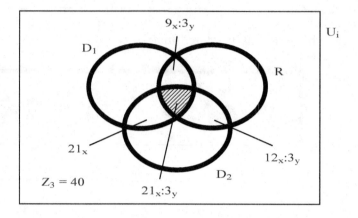

Figure 45.

For H:

$$n_r(\Phi_a) = \frac{n(D_1 \cup D_2)}{n(U_r)} = \frac{21}{24} = \frac{7}{8}$$

and

$$n_i(\Phi_a) = \frac{n(D_1 \cup D_2)}{n(U_i)} + \frac{n(D_1 \cup D_2)'}{n(U_i)} = \frac{21}{64} + \frac{43}{64} = 1.$$

Therefore

$$\widetilde{P} = \frac{7}{8}.$$

Multiple - Surrogate Theorem

The theorem of multiple surrogation is generally defined under the following conditions.

> *If the number of surrogate members of two or more numbers divisible by the same number are not zero, then the number of the surrogate members of their sum which is also divisible by the same divisor is the product of the number of surrogate members of the said two members.*

By definition, the multiple-surrogate theorem can be expressed mathematically as follows.

Let S_1, S_2, S_{n-1} and S_n be the surrogate sets for recipient sets R_1, R_2, R_{n-1} and R_n. Also let n_1, n_2 and n_n be numbers representing members of active distributive items and d the number representing the common active recipient(s). If

$$\frac{n_1}{d} = m_1 \ \text{with} \ n(S_1) = Z_1$$

$$\frac{n_2}{d} = m_2 \ \text{with} \ n(S_2) = Z_2$$

.

.

.

$$\frac{n_{n-1}}{d} = m_1 \ \text{with} \ n(S_{n-1}) = Z_{n-1}$$

then

$$n(S_n) = Z_n = Z_1 \cdot Z_2 \cdot \ ... \ \cdot Z_{n-1}$$

where Z_1, Z_2 and Z_{n-1} are not equal to zero and m_1, m_2 and m_{n-1} are real numbers.

Proof

Let the occurrence Z_1, Z_2, …, and Z_{n-1} in an ideal universal set, U_i be independent and that of Z_n be dependent on the rest (i.e. latter). Then, as shown in figure 46 below, the shaded

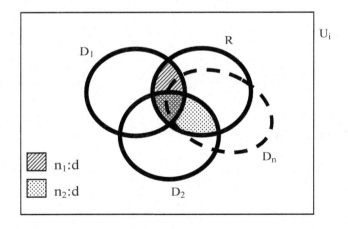

Figure 46

region represented by $D_1 \cap R$ is one of active distribution given by n_i:d. Also, in

$$D_2 \cap R \text{ we have } n_2\text{:d}$$

and generally in $D_n \cap R$ we have n_n:d. Note that: $n_1, n_2, ..., n_n$ and d are real numbers. The intersection of the nth shaded regions (i.e. the core) can be defined as

$$(D_1 \cap R) \cap (D_2 \cap R) \cap ... \cap (D_{n-1} \cap R) \cap (D_n \cap R).$$

This common region will be represented by the active distribution given by

$$(n_1 + n_2 + ... + n_{n-1} + n_n) : d.$$

By definition, Z_1 which results from n_1:d in $D_1 \cap R$ and Z_2 which results from n_2:d in $D_2 \cap R$ ... and Z_n which results from n_n:d in $D_n \cap R$ all occur in the region defined by

$$D_1' \cap D_2' \cap ... \cap D_{n-1}' \cap D_n' \cap R'.$$

Now, let n(U_i) be an arbitrary representation of the number of members in the ideal universal set. Then, in terms of probability

$$P(Z_n) \text{ in } D_1' \cap D_2' \cap R' = \{P(Z_1) \text{ from } D_1 \cap R\} \cap \{P(Z_2) \text{ in } D_2 \cap R\} \cap ... \cap \{P(Z_{n-1}) \text{ in } D_{n-1} \cap R\}$$

Let n(U_i)$_1$ be the number of members in the 1st ideal universal set, n(U_i)$_2$ that for the 2nd ideal universal set, ... and n(U_i)$_n$ that for the nth ideal universal set. Then

$$n(U_i)_n = n(U_i)_1 + n(U_i)_2 + ... + n(U_i)_{n-1}$$

Thus

$$\frac{Z_n}{n(U_i)_n} = \frac{Z_1}{n(U_i)_1} \cdot \frac{Z_2}{n(U_i)_2} \cdot \quad ... \quad \cdot \frac{Z_{n-2}}{n(U_i)_{n-2}} \cdot \frac{Z_{n-1}}{n(U_i)_{n-1}}$$

Multiplying through by respective n(U_i) gives

$$Z_n = Z_1 \cdot Z_2 \cdot \quad ... \quad \cdot Z_{n-2} \cdot Z_{n-1}$$

which is the result sort.

Multiple - Propensity Theorem

On the other hand, the quantitative analysis of the theorem of grouping the dividend results in a theorem referred to as the 'Multiple-Propensity Theorem'. It crystallizes from close comparison of the propensities derived.

The general advancement of the multiple-propensity theorem can be stated as follows.

> *Let the quotient probability ratios of the propensities $\tilde{P}_1$ and $\tilde{P}_2$ be respectively c_1/d_1 and c_2/d_2. Also, let these propensities pertain to the existence of recipient sets of two distributive processes A_1/B and A_2/B respectively. Then, the multiple propensity, $\tilde{P}_3$ of $(A_1+A_2)/B$ is given by*
>
> $$\tilde{P}_3 = \frac{c_1 + c_2}{d_1 + d_2 - 1}$$

Generally, it is observed that the propensity $\tilde{P}$ is given by

$$\tilde{P} = \frac{n(D)}{n(U_r)} = P(D)$$

where $P(D)$ is the probability of the distributive items.

Proof

Let the quotient probability ratio, P_o of the distributive processes be defined as follows

$$D_1 \cap R \text{ be } \frac{c_1}{d_1} \text{ where } n(U_i)_1 = d_1$$

$$D_2 \cap R \text{ be } \frac{c_2}{d_2} \text{ where } n(U_i)_2 = d_2$$

$$.$$
$$.$$
$$.$$

$$D_{n-1} \cap R \text{ be } \frac{c_{n-1}}{d_{n-1}} \text{ where } n(U_i)_{n-1} = d_{n-1}$$

$$D_n \cap R \text{ be } \frac{c_n}{d_n} \text{ where } n(U_i)_n = d_n$$

67

as shown in the figure 46 (under the proof for Multiple Surrogate). By definition, the denominator $n(U_i)$ of the simplest ratio, P_o (i.e. quotient probability ratio) includes the surrogate members under the condition where $P_o = \dfrac{Q}{n(U_i)}$ and Q is a number. But, in general

$$\tilde{P} = P(D) = \frac{n(D)}{n(U_r)}.$$

Therefore, it is effectively evident here that the denominator $n(U_r)$ precludes any surrogate member(s). The precluded surrogate member(s) is that of the 'simplest' number possible. This is due to the fact that P_o involves the simplest ratio possible. Also, since by definition multiple 'surrrogation' exists if neither of the products of the surrogate numbers are equal to zero, it can be logically implied that the simplest surrogating number possible is 1. Thus for multiple propensities

$$n(U_i)_n = n(U_i)_1 + n(U_i)_2 + \ldots + n(U_i)_{n-1} - 1$$
$$= d_1 + d_2 + \ldots + d_{n-1} - 1$$

where the negative sign signifies the preclusion of the simplest surrogate number. In reality, the simplest surrogating number is represented by the amount of elements in the distributor set (which is 1). By definition the distributor set, D is a singleton set due to its single membership. It must be borne in mind that propensity measures the natural tendency of the members of the recipient set R being a part of it. Since, there exists only one recipient set R, it is logically true that: $\tilde{P}_1, \tilde{P}_2, \tilde{P}_3, \cdots, \tilde{P}_{n-1}$ cannot all occur at the same time. Their occurrence must be one of a union. Hence

$$\tilde{P} = \tilde{P}_1 \cup \tilde{P}_2 \cup \tilde{P}_3 \cup \cdots \cup \tilde{P}_{n-1}$$

Therefore

$$\frac{c_n}{n(U_i)_n} = \frac{c_1}{n(U_i)_n} + \frac{c_2}{n(U_i)_n} + \ldots + \frac{c_{n-1}}{n(U_i)_n} = \frac{c_1 + c_2 + \ldots + c_{n-1}}{n(U_i)_n}$$

This gives

$$\frac{c_n}{n(U_i)_n} = \frac{c_1 + c_2 + \ldots + c_{n-1}}{d_1 + d_2 + \ldots + d_{n-1} - 1} = \tilde{P}$$

which is the result sort.

CHAPTER 4

INTER-OPERATIONAL RELATIONSHIPS

The attempt here is to demonstrate the basic relationships between the operations of division, sum, difference and product using the dividend and its divisor as the operands.

ANALYSIS OF ARITHMETIC OPERATIONS

Division-Sum Analysis

To determine this relationship, let a = Qx and b = x where *a* is the dividend operand, b the divisor operand, Q the quotient and x an equivalent divisor operand, then a/b = Q. Consequently, the relationship between a/b and a + b can be expressed as

$$a/b \propto (a+b)$$

which gives

$$a/b = K^+(a+b)$$

where K^+ is the '**summation or addition constant of proportionality**'. Substituting appropriately for the variables in the above equation we get

$$\frac{Qx}{x} \propto (Qx + x)$$

$$\frac{Qx}{x} = K^+(Qx + x)$$

$$Qx = K^+(Q+1)x^2$$

which can be expressed in the following quadratic form

$$K^+(Q+1)x^2 - Qx = 0$$

Factorizing gives

$$x\left[K^+(Q+1)x - Q\right] = 0$$

Finding the zero values yields

$$x = 0 \text{ or } K^+(Q+1)x - Q = 0$$

which implies

$$K^+ = \frac{Q}{(Q+1)x} \tag{6}$$

Observe that x (which is the divisor) also represents a zero value of the quadratic equation associated with the arithmetic operations analysis.

Example: Let a = 6 and b = 3. Then, $6/3 \propto (6+3)$ implies $6/3 = 9K^+$ and expressed as

$$K^+ = \frac{2}{9} = \left(\frac{2}{3}\right) \cdot \left(\frac{1}{3}\right).$$

By definition, the operation of division (i.e. a/b) is represented by $\underline{\underline{\Theta}}(a \mapsto b)$ where $\underline{\underline{\Theta}}$ is the division operator, *a* the dividend and b the divisor. Also, the operation of addition (i.e. a + b) is defined by $\underline{\underline{\Sigma}}(a \mapsto b)$ where $\underline{\underline{\Sigma}}$ is the addition operator. Thus, the direct proportionality constant between division and addition can be mathematically expressed as

$$\underline{\underline{\Theta}}(a \mapsto b) = K^+ \underline{\underline{\Sigma}}(a \mapsto b) \tag{7}$$

Division-Difference Analysis

Similarly, the relationship between division and subtraction operations of a dividend and a divisor is determined by

$$a/b \propto (a - b)$$

This by substitution leads to

$$\frac{Qx}{x} \propto (Qx - x)$$

$$\frac{Qx}{x} = K^- (Qx - x)$$

$$Qx = K^- (Q-1)x^2$$

where K^- is the '**difference or subtraction constant of proportionality**'. This leads to the following quadratic expression

$$K^- (Q-1)x^2 - Qx = 0.$$

Factorizing gives

$$x\left[K^- (Q-1)x - Q\right] = 0$$

By finding the zero values we get

$$x = 0 \text{ or } K^- (Q-1)x - Q = 0$$

which implies

$$K^- = \frac{Q}{(Q-1)x} \tag{8}$$

Example: Let a = 6 and b = 3. Then, $6/3 \propto (6-3)$ which implies $6/3 = 3K^-$ which boils down to

$$K^- = \frac{2}{3} = \left(\frac{2}{1}\right) \cdot \left(\frac{1}{3}\right).$$

By definition, the operation of division (i.e. a/b) is represented by $\underline{\underline{\Theta}}(a \mapsto b)$ where $\underline{\underline{\Theta}}$ is the division operator, a the dividend and b the divisor. Also, the operation of subtraction (i.e. a-b) is defined by $\underline{\underline{\Delta}}(a \mapsto b)$ where $\underline{\underline{\Delta}}$ is the subtraction operator. Thus, the direct proportionality constant between division and subtraction can be mathematically expressed as

$$\underline{\underline{\Theta}}(a \mapsto b) = K^- \underline{\underline{\Delta}}(a \mapsto b) \qquad (9)$$

Difference-Sum Analysis

To determine the relationship between the arithmetic operations of subtraction and addition, equation (6) is divided by equation (8) to give

$$\frac{K^+}{K^-} = \left[\frac{Q}{(Q+1)x} \right] \cdot \left[\frac{(Q-1)x}{Q} \right]$$

$$\frac{K^+}{K^-} = \frac{Q-1}{Q+1}.$$

Hence,

$$K^- = \left(\frac{Q+1}{Q-1} \right) \cdot K^+ \qquad (10)$$

Also, by dividing equation (7) by equation (9) we get

$$\frac{K^+}{K^-} \left(\frac{\underline{\underline{\Sigma}}(a \mapsto b)}{\underline{\underline{\Delta}}(a \mapsto b)} \right) = 1$$

Substituting for K^+/K^- and rearranging gives

$$\underline{\underline{\Delta}}(a \mapsto b) = \left(\frac{Q-1}{Q+1} \right) \cdot \underline{\underline{\Sigma}}(a \mapsto b) \qquad (11)$$

Division-Product Analysis

The relationship between division and multiplication is determined by letting
$$a/b \propto (ab).$$
By substituting for the variables a and b we obtain

71

$$\frac{Qx}{x} \propto \left(Qx^2\right)$$

$$\frac{Qx}{x} = K^\times\left(Qx^2\right)$$

$$Qx = K^\times\left(Qx^3\right)$$

where $K^\times$ is the '**product or multiplication constant of proportionality**'. Hence

$$K^\times = \frac{1}{x^2} \tag{12}$$

Example: Let a = 6 and b = 3. Then, $6/3 \propto 6(3)$ which implies $6/3 = 18K^\times$. Therefore

$$K^\times = \frac{1}{9} = \frac{1}{3^2}.$$

By definition, the operation of division (i.e. a/b) is represented by $\underline{\underline{\Theta}}(a \mapsto b)$ where $\underline{\underline{\Theta}}$ is the division operator, a the dividend and b the divisor. Also, the operation of multiplication (i.e. a x b) is defined by $\underline{\underline{\Pi}}(a \mapsto b)$ where $\underline{\underline{\Pi}}$ is the multiplication operator. Thus, the direct proportionality constant between division and multiplication can be mathematically expressed as

$$\underline{\underline{\Theta}}(a \mapsto b) = K^\times \underline{\underline{\Pi}}(a \mapsto b) \tag{13}$$

Product-Sum Analysis

To determine the relationship between the arithmetic operations of multiplication and addition, equation (7) is divided by equation (13) to give

$$\frac{K^+}{K^\times}\left(\frac{\underline{\underline{\Sigma}}(a \mapsto b)}{\underline{\underline{\Pi}}(a \mapsto b)}\right) = 1$$

which can be expressed as

$$\frac{K^\times}{K^+} = \left(\frac{\underline{\underline{\Sigma}}(a \mapsto b)}{\underline{\underline{\Pi}}(a \mapsto b)}\right)$$

Substituting for the ratio $K^\times / K^+$ using equations (12) and (6) respectively, we get

$$\frac{K^\times}{K^+} = \frac{Q+1}{xQ} = \left(\frac{\underline{\underline{\Sigma}}(a \mapsto b)}{\underline{\underline{\Pi}}(a \mapsto b)}\right).$$

This equation can be expressed in two forms as

$$K^{\times} = \left(\frac{Q+1}{a}\right)K^{+}$$

(14)

where by definition, the product of the divisor and quotient equals the dividend (i.e. xQ = a). Alternatively, the second expressed equation is given as

$$\underline{\underline{\Pi}}(a \mapsto b) = \left(\frac{a}{Q+1}\right)\underline{\underline{\Sigma}}(a \mapsto b)$$

(15)

Product-Difference Analysis

The relationship between the arithmetic operations of multiplication and subtraction is derived by dividing equation (9) by equation (13). This gives

$$\underline{\underline{\Delta}}(a \mapsto b) = \left(\frac{Q-1}{a}\right)\underline{\underline{\Pi}}(a \mapsto b)$$

(16)

Division and Product-Difference Ratio Analysis

The relationship between the arithmetic operations of division and the ratio of multiplication to subtraction can be derived by multiplying both sides of equation (16) by 1/b. This gives

$$\underline{\underline{\Theta}}(a \mapsto b) = \left(\frac{Q-1}{a}\right)\left(\frac{\underline{\underline{\Pi}}(a \mapsto b)}{\underline{\underline{\Delta}}(a \mapsto b)}\right)$$

(17)

Division and Product-Sum Ratio Analysis

The relationship between the arithmetic operations of division and the ratio of multiplication to addition can be derived by multiplying both sides of equation (15) by 1/b. This gives

$$\underline{\underline{\Theta}}(a \mapsto b) = \left(\frac{Q-1}{b}\right)\left(\frac{\underline{\underline{\Pi}}(a \mapsto b)}{\underline{\underline{\Sigma}}(a \mapsto b)}\right)$$

(18)

73

DONOR SURROGATE AND UNIT DONOR DISTRIBUTION PRINCIPLE

Let us evaluate 0/1 using equation (11). First, it is seen that if a/b =0/1 then the operation of subtraction is given by

$$\underline{\underline{\Delta}}(a \mapsto b) = a - b = 0 - 1 = -1$$

and that of the operation of addition by

$$\underline{\underline{\Sigma}}(a \mapsto b) = a + b = 0 + 1 = 1.$$

Hence applying equation (11) to solve for the quotient

$$-1 = \left(\frac{Q-1}{Q+1}\right)$$
$$-Q - 1 = Q - 1$$
$$0 = 2Q - 1 + 1$$
$$2Q = 0$$
$$Q = \frac{0}{2} = 0$$

which is logically and realistically true. The only existing single recipient will get nothing if nothing is shared.

However, if 0/0 is evaluated using equation (11), the situation becomes different. Here, applying equation (11) gives

$$0 = \left(\frac{Q-1}{Q+1}\right) \cdot 0$$

which implies 0 = 0. But since we are interested in finding the value of the quotient, Q under the desired or given condition, equation (11) is rearranged as follows

$$\frac{\underline{\underline{\Delta}}(a \mapsto b)}{\underline{\underline{\Sigma}}(a \mapsto b)} = \frac{Q-1}{Q+1}.$$

Substituting for $\underline{\underline{\Delta}}(a \mapsto b) = a - b = 0 - 0 = 0$ and $\underline{\underline{\Sigma}}(a \mapsto b) = a + b = 0 + 0 = 0$ we get

$$\frac{0}{0} = \frac{Q-1}{Q+1} \tag{19}$$

Observe that the LHS of the equation represents the distributive situation whose quotient is sort. Thus, if we let $Q_- = -1$ and $Q_+ = 1$ be partial quotients. Then the equivalent expression of equation (19) is given as

$$\frac{0}{0} = \frac{Q-1}{Q+1} \equiv \frac{Q_+ - 1}{Q_- + 1} = \frac{0}{0}$$

which is true. On the other hand

$$\textit{Effective Quotient} = \frac{0}{0} = \frac{Q-1}{Q+1} \equiv \frac{Q_- - 1}{Q_+ + 1} = \frac{-1-1}{1+1} = \frac{-2}{2} = -1.$$

The result here implies each surrogate member received -1 item. What does it logically and realistically mean? Since the recipient set is non-existent, its member size must be zero. Also although an empty set, the distributor set D exists because a new form of receiving set exists. It is called the '**donor-surrogate set**', S_0. The donor-surrogate set is disjointed from the distributor set. Consequently, the real situation of 0/0 in the real universal set (i.e. U_r) scenario as shown in figure 47 below has an equivalent situation in the ideal universal set (i.e. U_i) scenario.

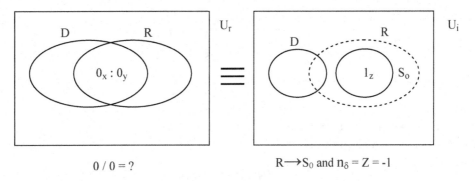

Figure 47

The '**unit donor distribution principle**' which is the resulting realistic meaning for $n(S) = 1_z$ states that

If there exists a distributor with no distributive item(s) to work with and there exists no recipient, then there exists a natural tendency for one surrogate person (out of the population in the universal set) to be willing to give out an item for the purpose of sharing.

As such, the '**donor quotient**', n_δ is the auto-donated item by the unit surrogate recipient. Consequently

$$\frac{0}{0} = n_\delta = -1.$$

Also, the unit donor distribution principle further suggests that,

Any distributional process has a tendency of taking place.

75

CONCEPT OF SET RESONANCE

The evaluation of 1/0, a case of division by zero, using equation (11) reveals a very unique situation. That is if

$$\frac{a}{b} = \frac{1}{0}$$

then the operation of subtraction is given by

$$\underline{\underline{\Delta}}(a \mapsto b) = a - b = 1 - 0 = 1$$

and that of the operation of addition by

$$\underline{\underline{\Sigma}}(a \mapsto b) = a + b = 1 + 0 = 1.$$

Hence applying equation (11) to solve for the quotient we get

$$1 = \left(\frac{Q-1}{Q+1}\right)$$

This can be expressed as

$$Q + 1 = Q - 1$$

Let the partial quotient associated with the LHS of the equation immediately above be Q_- and that associated with the RHS be Q_+. Then

$$Q_- + 1 = Q_+ - 1 \tag{20}$$

where Q_- is the opposite of Q_+ or $-Q_- = Q_+$ and $-Q_+ = Q_-$. Grouping all Qs of equation (20) to the RHS we obtain

$$2 = Q_+ - Q_-.$$

Since $-Q_- = Q_+$ we get $2 = 2Q_+$ which implies $Q_+ = 1$. On the other hand, if we group all Qs in equation (20) to the LHS we obtain

$$Q_- - Q_+ = -2.$$

Since $-Q_+ = Q_-$ we can write $2Q_- = -2$ which implies $Q_- = -1$. Hence

$$\frac{a}{b} = \frac{1}{0} = Q_\pi = \sqrt{1} = \pm 1$$

where $Q_+ = 1$, $Q_- = -1$ and Q_π is referred to as the '**periodic quotient**'. The value of periodic quotient indicates that an oscillation exists between a donor surrogate set, S_0 and an interested or interacting surrogate set, S. Consequently, a periodic surrogate set's oscillation occurs between S and S_0. This process creates what is dubbed the '**standing set surrogate waves**'. In general,

$$\frac{m}{0} = Q_\pi = \sqrt{1} = \pm 1$$

where m is a real number

and

$$\frac{0}{0} = Q_\delta = -1.$$

The oscillatory nature of the periodic quotient, Q_π would be shown later under '*The Logarithm of Zero*', as equal to the '**unsigned zero**' or '**mean zero**', $\langle \tilde{0} \rangle$ which is technically equivalent to the conventional zero, 0. It must be noted though that the inverse of a division process which is multiplication cannot be invoked here due to the periodic or oscillatory nature of the process.

To further identify the exact relationship between the recipient set, R and those of S and S_o, use is made of equations (15) and (16) respectively. Using equation (15), it is observed that if

$$\frac{a}{b} = \frac{1}{0}$$

Then

$$\underline{\underline{\Sigma}}(a \mapsto b) = 1,\ \underline{\underline{\Pi}}(a \mapsto b) = 0,\ \underline{\underline{\Delta}}(a \mapsto b) = 1 \text{ and } a = 1.$$

Therefore

$$\frac{1}{0} = Q + 1.$$

From the previous results for 1/0 where $Q_\pi = \pm 1$, the implication is that

$$Q = 0 \text{ for } Q_+ = 1.$$

On the other hand, using equation (16) to analyze the ratio 1/0 we get

$$\frac{1}{0} = Q - 1.$$

This implies

$$Q = 0 \text{ for } Q_- = 1.$$

The above situation validly occurs when n = 0 for all cases of $Q_\pi = \pm 1$ or $Q_+ = 1$ and $Q_- = -1$ identified. It shows that a recipient set of in a division- by- zero scenario is bounded by the sets S and S_o. Hence this recipient set is called the '**mean recipient set**', $\overline{R}$. The periodic surrogate set's oscillation about the set $\overline{R}$ is illustrated in the form of a Venn diagram shown in figure 48 below. Also, a cross-sectional view of the resulting standing set's surrogate waves is shown

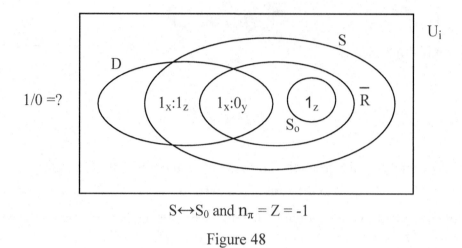

$$S \leftrightarrow S_0 \text{ and } n_\pi = Z = -1$$

Figure 48

in figure 49 below. Notice that S_0, $\overline{R}$ and S all lie on the 0-0 line. Since the distributor normally donates and not a surrogate, the quotient scale at S_0 has a negative sense (downward direction in figure 49). On the other hand, because no donation occurs with S, the quotient scale at S has a positive sense (upward direction). However, the position of the mean recipient set is neutral. Unlike S_0 and S, $\overline{R}$ neither rises nor falls.

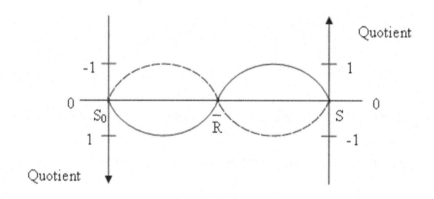

Figure 49

By definition, resonance occurs when one object vibrating at the same natural frequency of a second object forces that second object into vibrational motion. Thus, the forced frequency of S_0 affects that of S into having the same magnitude of frequency (i.e. the number of element or member size). Consequently, not only is there an existing standing wave but also a kind of '**set resonance**' is also setup. The combination of these two phenomena will be called '**standing set surrogate resonance**'.

The implication of the results here is that under equity distributions, if there is no recipient in the light of an existing distributive item, a recipient would naturally be sort with equal intensity in an even manner for donation until one is found. The propensity of finding such a recipient (mean) will be 50%.

FUNDAMENTAL PRINCIPLES OF EQUITY DISTRIBUTIONS

The general principles governing equity distributions can be summed up as follows.

Let $n(S)$ be the number of members of an interested or interacting surrogate set, S and $n(S_o)$ be the number of members of a donor surrogate set, S_o in any equity distribution process, then

1. *$n(S)$ is a constant provided $n(D)$ and $n(R)$ are non-zero quantities and remain the same under any distributive variation or varying distributive conditions.*

2. *$n(S)$ and $n(S_o)$ are both constants equal to unity (i.e. 1) provided $n(R)$ is equal to 0 and $n(D)$ is non-zero under varying distributive conditions. S and So are said to be singleton sets.*

3. *$n(S_o)$ is a constant equal to unity provided both $n(D)$ and $n(R)$ are all equal to zero.*

It must be noted that the three principles stated above are fundamental to the processes of equity distributions.

CHAPTER 5

EXCOGITATING UNKNOWN MATHEMATICAL EXPRESSIONS

An algebraic expression obtained within the context of limits is said to be an indeterminate form if there exist no information to allow the limit of the algebraic expression to be determined. Indeterminate forms found in calculus and other branches of mathematical analysis include the following: 0^0, $0/0$, 1^∞, $\infty - \infty$, ∞/∞, $0\times\infty$, and ∞^0. [6] On the other hand, while the following expressions: $\infty - \infty$, $(-1)^{\pm\infty}$, $0(-\infty)$ and $\pm\infty/\pm\infty$ are said to be undefined in all contexts others such as $x/0$, 0^0, ∞^0, 1^∞, and $0\times\infty$ are not defined in all contexts. [7] The lack of reasonable mathematical expression that has meaningful, unambiguous and sensible value is said to be undefined.

Frequently, infinity is used to define a limit. In extended real number system, it is used as a value for which some arithmetic operations may be performed. These are positive infinity (∞) which is greater than all other extended real numbers, and negative infinity ($-\infty$) which is lesser than all other extended real numbers.

Interestingly, most branches of mathematics disagree with the question of 0^0 because their different approaches do not result in the same answer. While Cauchy viewed 0^0 as undefined, it is frequently considered in modern textbooks as equal to 1. [8] The inconsistency in using the limits approach can not be over emphasized. For example, as x decreases to 0, the functions 0^x and x^0 have different limiting values. Ronald Graham et al opt for $x^0 = 1$ for all x on the basis of the importance of the validity of the binomial theorem when $x = 0$, $y = 0$, and/or $x = -y$.

Unlike the limits approach that have been subjectively used to determine the values of the said algebraic expressions, a new concept called '**logarithmic quotient**' will be objectively used to realistically and logically determine the values of some of the indeterminate, undefined algebraic and infinity expressions.

LOGARITHM OF A NEGATIVE NUMBER

Logarithm is a concept that can be shown to be based on division. For example

$$\log_2 8 = 3$$

which by definition gives

$$8 = 2^3 = 2 \cdot (2 \times 2)$$

This can be expressed in terms of division as

$$\frac{8}{2} = (2 \times 2) = 4$$

Let the given number y, the base b and the exponent or power x all be positive real numbers. If,

$$\log_b y = x \quad then$$

$$\frac{y}{b} = b^{x-1}$$

It is observed that with the logarithm of a positive number which is generally expressed as

$$\log_b y = x$$

1. the given number and the base are all positive.
2. the given number and the exponent are not constants or the given number varies with the exponent.
3. the base is a constant.

On the other hand, the logarithm of a negative number can be exemplified as follows

$$\log_{-8} -2 = \frac{1}{3}$$

$$\Rightarrow -2 = -8^{\frac{1}{3}} = \sqrt[3]{-8}$$

Consequently, with the logarithm of a negative number which is expressed generally as

$$\log_{-b}(-y) = \phi x$$

where ϕ is an odd number and x is a real number, it is observed that the

1. given number and the base are all negative.
2. given number and the base vary or are not constants.
3. exponent term is a constant.

In general, in a positive logarithmic function y maps into x (i.e. $y \to x$) while in a negative logarithmic function a negative y maps into a negative b (i.e. $-y \to -b$).

Finding the Value of X.

The equation for the general logarithm of a negative number is expressed as an exponent as follows

$$-y = (-b)^{\phi x} \tag{21}$$

For example, if x = 1 then $-8 = (-2)^{3x}$ is equivalent to $8 = 2^{3x}$. Taking $\log_{10}$ of both sides of the latter equation we get

$$\log_{10} 8 = \log_{10} 2^{3x} = 3x \log_{10} 2$$

81

By making x the subject of the equation we get

$$3x = \frac{\log_{10} 8}{\log_{10} 2}$$

which gives

$$x = \left(\frac{1}{3}\right)\frac{\log_{10} 8}{\log_{10} 2} = 1.$$

Thus, equation (21) can be evaluated for x by relegating the negative signs. This gives

$$y = b^{\phi x}$$

Taking log of both sides results in

$$\log y = \phi x \log b$$
$$\phi x = \frac{\log y}{\log b}$$

This finally gives

$$x = \left(\frac{1}{\phi}\right)\frac{\log y}{\log b}$$

Example 1. Find $\log_{-10}(-10)$ using $\phi = 3$.

Solution: Let,

$$\log_{-10}(-10) = \phi x = 3x$$

This gives

$$-10 = (-10)^{3x}$$

which can be equivalently expressed as $10 = 10^{3x}$. Equating the exponents we have $1 = 3x$ which results in $x = \frac{1}{3}$.

Example 2. Find $\log_{-10}(-7)$ using $\phi = 3$

Solution: Let

$$\log_{-10}(-7) = \phi x = 3x$$

which can be written as

$$-7 = (-10)^{3x}$$

This is equivalently expressed as

$$7 = 10^{3x}$$

Taking log10 of both sides yields

$$\log_{10} 7 = 3x \log_{10} 10$$

$$3x = \frac{\log_{10} 7}{\log_{10} 10}$$

$$x = \frac{1}{3}\left[\frac{\log_{10} 7}{\log_{10} 10}\right] = 0.281699347$$

Hence

$$\log_{-10} - 7 = \phi\, x = 3(0.281699347) = 0.845098041 \qquad\qquad (22)$$

In general,

$$\log_{-b} - y = \phi\, x = \frac{\log_n y}{\log_n b}$$

where n is a positive real number.

The problem lingering here is how to show that the result of 3ϕ in equation (22) is actually equal to the logarithm of a negative number. First, if one applies the basic principle of logarithm it implies

$$-7 = (-10)^{0.845098041}$$

which is an error because

$$7 = 10^{0.845098041}$$

This situation is surmounted by expressing ϕx as a difference between a whole number and a decimal. In the situation above

$$\phi\, x = 3x = 0.845098041 = 1 - 0.154901959$$

Hence

$$-7 = (-10)^{1-0.154901959}$$

Using laws of indices

$$-7 = \frac{(-10)}{(-10)^{0.154901959}}$$

For the RHS of the equation above to be a negative number either the numerator is negative or the denominator is negative but not both. Since the numerator is negative and the denominator is still undefined, it is appropriate to let the denominator be equivalent to a positive number. Hence, by definition

$$-7 = \frac{(-10)}{10^{0.154901959}} = \frac{-10}{1.428571425} = -7.$$

In a similar manner

$$\log_{-10}(-20) = \phi\, x = \frac{\log_{10} 20}{\log_{10} 10} = \log_{10} 20 = 1.301029996.$$

Expressing ϕx in the form of a whole number and a decimal gives

$$\phi\, x = 1 + 0.301029996.$$

Therefore, from the logarithm of a negative number

$$-20 = (-10)^{1+0.301029996} = (-10) \cdot (-10)^{0.301029996}.$$

But, by definition

$$(-10)^{0.301029996} \equiv 10^{0.301029996}$$

Thus

$$-20 = (-10) \cdot 10^{0.301029996}$$
$$= (-10) \cdot (2.000000002)$$
$$= -20.$$

Theorem of Negative Logarithm

Formally, the theorem of negative logarithm can be stated as follows.

Let b, y and X' be real numbers. Then

1. *The logarithm of a negative number is defined as,*

$$\log_{-b}(-y) = X'$$

provided $X' = \dfrac{\log_n y}{\log_n b}$ *and n is a positive real number.*

2. *The exponent is expressed as,*

$$X' = M \pm m$$

provided $M \neq 0$ *and* $0 < m < 1$ *where M is a positive integer and m is a decimal part.*

3. *If* $X' = M + m$ *then*

$$-y = (-b)^M \cdot (b)^m$$

provided $(-b)^M \equiv (b)^m.$

4. *If* $X' = M - m$ *then*

$$-y = \dfrac{(-b)^M}{(b)^m}$$

provided $(-b)^M \equiv (b)^m.$

QUOTIENT NATURE OF LOGARITHMS

Let,

$$\log_b y = x$$

then, by simply expressing the LHS of the equation above in another base say n using the following law of logarithm

$$\log_a b = \frac{\log_c b}{\log_c a}.$$

we get

$$x = \frac{\log_n y}{\log_n b}.$$

From the above result, it is seen that x represents a quotient which is appropriately dubbed '**logarithmic quotient**'. Consequently, logarithm can be analyzed under an '**obelus space**' (which is a set of functions with common division operation) just as the arithmetic operation of division.

LOGARITHM OF ZERO

Let,

$$\log_{-10} 0 = X' = \phi \left(\frac{N}{0} \right) \tag{23 - 1}$$

where X' is a division by a conventional zero, the factor ϕ an arbitrary odd number and N a real number. Then, by the definition of negative logarithm we have

$$\phi \left(\frac{N}{0} \right) = \frac{\log_{10} 0}{\log_{10} 10} = \log_{10} 0 \tag{23 -2}$$

since from contemporary laws of logarithm,

$$\log_b b = 1.$$

This implies

$$\phi = \left(\frac{0}{N} \right) \log_{10} 0$$

Therefore, from equation (23 - 1)

$$0 = (-10)^{\left(\frac{0}{N}\right)(\log_{10} 0)\left(\frac{N}{0}\right)}$$

$$0 = (-10)^{\left(\frac{0}{N}\right)\left(\frac{N}{0}\right)} + (-10)^{\log_{10} 0}$$

$$0 = (-10)^{0\left(\frac{N}{0}\right)} + (-10)^{\log_{10} 0}$$

$$0 = (-10)^{0} + (-10)^{\log_{10} 0}$$

$$0 = 1 + (-10)^{\log_{10} 0}$$

Rearranging we get

$$(-10)^{\log_{10} 0} = -1$$

Taking log 10 of both sides, we can write

$$\log_{10} 0 \left[\log_{10}(-10)\right] = \log_{10}(-1)$$

which boils down to

$$\log_{10} 0 = \frac{\log_{10}(-1)}{\log_{10}(-10)} \tag{24}$$

Also, from the laws of logarithm

$$\log_a b = \frac{\log_c b}{\log_c a}$$

Therefore

$$\log_{10}(-1) = \frac{\log_{-10}(-1)}{\log_{-10} 10} \quad \text{and} \quad \log_{10}(-10) = \frac{\log_{-10}(-10)}{\log_{-10} 10}.$$

Hence, by substituting the above results into equation (24) we get

$$\log_{10} 0 = \frac{\log_{-10}(-1)}{\log_{-10} 10} \cdot \frac{\log_{-10} 10}{\log_{-10}(-10)}$$

which can also be expressed as

$$\log_{10} 0 = \frac{\log_{-10}(-1)}{\log_{-10}(-10)} \tag{25}$$

To find the value of X' for each logarithm in equation (25), we have the following.

For $\log_{-10}(-1)$:
$$X_1' = \frac{\log_{10} 1}{\log_{10} 10} = \log_{10} 1 = 0.$$

Therefore,
$$X_1' = 1 - 1 = 0.$$

This implies
$$-1 = \frac{(-10)^1}{10^1} = \frac{-10}{10} = -1.$$

For $\log_{-10}(-10)$:

$$X'_2 = \frac{\log_{10} 10}{\log_{10} 10} = 1.$$

Therefore,

$$X'_2 = 1 + 0 = 1$$

which implies

$$-10 = (-10)^1 \cdot (10)^0 = -10.$$

Or

$$X'_2 = 1 - 0 = 1$$

which boils down to

$$-10 = \frac{(-10)^1}{10^0} = -10.$$

The Signed Distributive Zeroes (SDZ)

The numbers making up the SDZ are the '**positive zero**' and '**negative zero**'. The proofs for their definitions follow.

By the definition of negative logarithm, using equations (23 - 1), (25) and values of X' above

$$X' = \phi\left(\frac{N}{0}\right) = \log_{10} 0 = \frac{X'_1}{X'_2} = \frac{0}{1} = 0.$$

It implies, the arbitrary odd number is transformed as $\phi = 0$. This is realistically and logically true because ϕ which represents the factor for maintaining a negative base when raise to the power X' is no longer needed and so it becomes null and void. The reason for

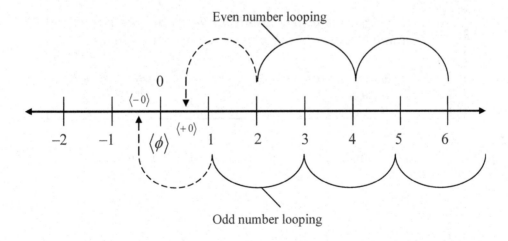

Figure 50 a. Identification of zeroth number field's sense-based even and odd number on a real number line with the aid of an even and odd number looping.

this is that the number y whose logarithm is sort, is equal to a conventional zero (see equation (23 - 1). The nullification of the arbitrary odd number cannot represent a conventional zero but a new form of zero dubbed '**negative zero**' and represented by the symbol $\langle-0\rangle$. This can be deduced with the aid of figure 50a below. By interpolating the odd number looping, the immediate odd number less than 1 is $\langle-0\rangle$. Hence, it is called the '**zeroth odd number**'. On the other hand, that for the even number looping is $\langle+0\rangle$ which is dubbed '**positive zero**'. It is also called the '**zeroth even number**'. Observe that a vertical line through the number 0 on the number line will represent the line of numeric symmetry. It demarcates the positive and negative numbers on the number line. Consequently from equation (23 - 1), it can be concluded that

$$\log_{10} 0 = X' = \langle-0\rangle\left(\frac{N}{0}\right) = \langle-0\rangle$$

since $\langle-0\rangle$ is a form of zero. This implies that

$$\boxed{\log_{10} 0 = \langle-0\rangle} \qquad (26 - 1)$$

From equations (26 - 1), (24) and (25) we can write

$$\langle-0\rangle = \frac{\log_{-10}-1}{\log_{-10}-10} = \frac{\log_{10}-1}{\log_{10}-10}$$

In principle, a '**universal zeroth law**' can be states as follows:

.

> 1. *The product of any zeroth number and any non-zero number is always equal to the same zeroth number.*
>
> 2. *Any number raised to the power of any zeroth number is equal to one.*

Consequently, the result from equations (26 - 1), (24) and (25) stated above can be expressed as

$$\log_{-10}-1 = \langle-0\rangle\log_{-10}-10 \Rightarrow \log_{-10}-1 = \langle-0\rangle$$

or

$$\log_{10}-1 = \langle-0\rangle\log_{10}-10 \Rightarrow \log_{10}-1 = \langle-0\rangle$$

which translate into

$$\boxed{\begin{array}{c} -1 = 10^{\langle -0 \rangle} \\ or \\ -1 = -10^{\langle -0 \rangle} \end{array}} \qquad (26-2)$$

The value -1 at the LHS of equations in $(26-2)$ is technically not a conventional -1 but related to a form of integer that I call a '**neutral unit integer**' and denoted by the symbol $\langle \tilde{1} \rangle$. Later on, its nature will be appropriately explained. Thus, it can state generally that

$$-1 = +b^{\langle -0 \rangle} = -b^{\langle -0 \rangle} \equiv \langle \tilde{1} \rangle \qquad (26-3)$$

where $+b$ and $-b$ represent positive and negative real numbers respectively.

Using equation $(26-2)$ to find the value of the 'positive zero' we have

$$-1 = -10^{\langle -0 \rangle}$$

But

$$-10 = 10^0 \cdot (-10)$$

Therefore

$$-1 = \left[10^0 \cdot (-10) \right]^{\langle -0 \rangle}$$

$$-1 = 10^{0(\langle -0 \rangle)} \cdot (-10)^{\langle -0 \rangle} \qquad (26-4)$$

Since, by definition

$$\boxed{\langle -0 \rangle = -\langle +0 \rangle}$$

We have

$$10^{0(\langle -0 \rangle)} = (10)^{0(-\langle +0 \rangle)} = 10^{0-\langle +0 \rangle}$$

which can be written in accordance with the laws of indices as

$$10^{0(\langle -0 \rangle)} = 10^{0-\langle +0 \rangle} = \frac{10^0}{10^{\langle +0 \rangle}}$$

Therefore from equation $(26-4)$

$$-1 = \frac{10^0}{10^{\langle +0 \rangle}} \cdot (-10)^{\langle =0 \rangle}$$

$$-1 = \frac{(-10)^{\langle -0 \rangle}}{10^{\langle +0 \rangle}}$$

By cross multiplying we get

$$(-1) \cdot 10^{\langle +0 \rangle} = (-10)^{\langle -0 \rangle}$$

Therefore from equation (26 – 2), it can be concluded that

$$10^{\langle +0 \rangle} = \frac{(-10)^{\langle -0 \rangle}}{-1} = -(-10)^{\langle -0 \rangle} = -(-1) = 1$$

Hence

$$1 = +b^{\langle +0 \rangle} = -b^{\langle +0 \rangle} \equiv \langle \widetilde{1} \rangle \qquad (26\text{ - }5)$$

Consequently, it can be concluded from equations (26 - 3) and (26 – 5) that

$$\boxed{\log_{-10} 0 = \langle +0 \rangle} \qquad (26\text{ - }6)$$

From the equation below

$$r^0 = 1$$

where r is any number, it can be deduced from equation (26 – 5) that the value 1 at the beginning of equation (26 - 5) is not a conventional number. This is so, since

$$\langle +0 \rangle \neq 0$$

The value 1 in equation (26 – 5) is also related to the said '**neutral unit integer**' which will be analyzed later.

The Neutral Distributive Zero (NDZ)

This type of zero is also called '**mean zero**'. It consists of a positive and negative zero co-existing together. Below is a proof to its definition.

From equation (26 – 3)

$$-1 = (-b)^{\langle -0 \rangle}$$

But, minus b can be expressed as

$$-b = (-b) \cdot 1$$

So, we can write

$$-1 = [(-b) \cdot 1]^{\langle -0 \rangle}$$
$$-1 = (-b)^{\langle -0 \rangle} \cdot 1^{\langle -0 \rangle}$$

Therefore

$$\frac{-1}{(-b)^{\langle -0 \rangle}} = 1^{\langle -0 \rangle} \qquad (26 - 27)$$

But, from equation (26 – 3)

$$\frac{-1}{-1} = 1^{\langle -0 \rangle} = 1$$

Similarly, if we use from equation (26 – 5)

$$1 = (+b)^{\langle +0 \rangle}$$

we will end up with

$$\frac{1}{1} = 1^{\langle -0 \rangle} = 1$$

Therefore, in general

$$(1)^{\langle +0 \rangle} = (1)^{\langle -0 \rangle} = 1.$$

On the other hand, from equation (26 – 27) we can write

$$(-1)(-1)^{-\langle -0 \rangle} = 1^{\langle -0 \rangle}$$
$$(-1)(-1)^{\langle +0 \rangle} = 1^{\langle -0 \rangle}$$
$$(-1)^{\langle +0 \rangle} = (-1)\left(1^{\langle -0 \rangle}\right)$$

By definition, $\langle +0 \rangle$ is a 'zeroth even number' therefore the LHS of the above equation will be negative in magnitude. Also, $\langle -0 \rangle$ is a 'zeroth odd number'. Hence, the factor $1^{\langle -0 \rangle}$ will be positive in value and so the RHS will end up being a negative value. Consequently, since $\langle +0 \rangle$ has a positive numeric sense and $\langle -0 \rangle$ has a negative numeric sense, one can write in terms of magnitude that,

$$\left| \langle +0 \rangle \right| = \left| \langle -0 \rangle \right|$$

Thus, by virtue of numeric sense it can safely be state that

$$\langle \tilde{0} \rangle = \langle +0 \rangle + \langle -0 \rangle$$

where $\langle \tilde{0} \rangle$ which is the 'mean zero' has no numeric sense.

Types of Signed Non-Distributive Zeroes (SNDZ)

This involves division by signed distributive zeroes. There are two types namely '**positive SNDZ**' and '**negative SNDZ**'. The proofs for the types of signed non-distributive zeroes are provided below.

Applying the conventional laws of logarithm, we have

$$\log_{10} 0 = \frac{\log_{-10} 0}{\log_{-10} 10}$$

91

which can be written as

$$\log_{-10} 10 = \frac{\log_{-10} 0}{\log_{10} 0} = \frac{\langle +0 \rangle}{\langle -0 \rangle}$$

Thus, the '**positive SNDZ**' which is based on a positive zero distribution can be expressed mathematically as

$$\boxed{\log_{-10} 10 = \frac{\langle +0 \rangle}{\langle -0 \rangle}} \qquad (26 - 28)$$

where the RHS term represents the 'positive SNDZ'. Alternatively, by expressing the LHS of equation (26 - 28) as follows

$$\log_{-10} 10 = \frac{\log_{10} 10}{\log_{10} -10} = \frac{1}{\log_{10} -10}$$

But

$$\log_{10} -10 = \frac{1}{\log_{-10} 10}$$

Therefore, the '**negative SNDZ**' which is based on a negative zero distribution can be expressed mathematically as

$$\boxed{\log_{10} -10 = \frac{\langle -0 \rangle}{\langle +0 \rangle}} \qquad (26 - 29)$$

where the RHS term represents the 'negative SNDZ'.

The Unsigned Distributive Zero (UDZ) and Signed Distributive Zero (SDZ)

By definition, from equation (26 – 29) we can write

$$\langle -0 \rangle \log_{-10} 10 = \langle +0 \rangle$$

$$\log_{-10} 10^{\langle -0 \rangle} = \langle +0 \rangle$$

$$10^{\langle -0 \rangle} = -10^{\langle +0 \rangle}$$

which gives

$$\boxed{\begin{array}{c} 10^{\langle -0 \rangle} + 10^{\langle +0 \rangle} = 0 \\ \textit{provided} \\ \log_{-10} 0 = \langle +0 \rangle \end{array}} \qquad (26 - 30)$$

92

Proof

By definition,

$$\log_{-10} 0 = \langle +0 \rangle$$

$$0 = (-10)^{\langle +0 \rangle}$$

Substituting the above result into equation (26 – 30), we get

$$10^{\langle -0 \rangle} + 10^{\langle +0 \rangle} = (-10)^{\langle +0 \rangle}$$

But, the positive zero is a zeroth odd number. Therefore the RHS term of the above equation will be positive in value. Hence,

$$10^{\langle -0 \rangle} + 10^{\langle +0 \rangle} = 10^{\langle +0 \rangle}$$

$$10^{\langle -0 \rangle} + 10^{\langle +0 \rangle} - 10^{\langle +0 \rangle} = 0$$

$$10^{\langle -0 \rangle} = 0$$

$$\underline{\underline{\log_{10} 0 = \langle -0 \rangle}}$$

which is a true result.

The Unsigned Non-Distributive Zero (UNDZ) and Void Zero

A '**void zero**' denoted by $\langle \phi \rangle$, results when the positive and negative zeroes of a mean zero nullifies each other in terms of numeric senses. Its relationship with an unsigned non-distributive zero will be determined here.

From the conventional laws of logarithm, the following UNDZ can be expressed as

$$\log_b \left(\frac{0}{0} \right) = \log_b 0 - \log_b 0$$

$$= \langle +0 \rangle - \langle +0 \rangle$$

$$= \langle -0 \rangle - \langle -0 \rangle$$

$$= \langle \phi \rangle$$

Here the RHS terms have a neutralizing effect. Alternatively,

$$\log_b \left(\frac{0}{0} \right) = \langle -0 \rangle - \langle -0 \rangle$$

$$= \langle -0 \rangle + \langle +0 \rangle$$

$$= \langle \tilde{0} \rangle$$

93

which implies the positive and negative zeroes terms at the RHS are co-existing.

Therefore, it can be generalized that,

$$\left(\frac{0}{0}\right) = b^{\langle \tilde{0} \rangle} \equiv b^{\langle \phi \rangle} \qquad (26-31)$$

where the LHS term is the UNDZ.

The table below gives a comparison between the common names and the distributive names of the elements of the zeroth number field.

NUMERICAL SYMBOLS	COMMON NAMES	DISTRIBUTIVE NAMES
0	Conventional zero	Unsigned distributive zero (UDZ)
$\langle -0 \rangle$	Negative zero	Negatively signed distributive zero (NSDZ)
$\langle +0 \rangle$	Positive zero	Positively signed distributive zero (PSDZ)
$\langle \tilde{0} \rangle$ or $\langle \pm 0 \rangle$	Mean zero	Neutral distributive zero (NDZ)
$\langle \phi \rangle$	Void zero or Null zero	Zilch distributive zero (ZDZ)
$\left(\frac{0}{0}\right)$	Ratio of conventional zero to conventional zero	Unsigned non-distributive zero (UNDZ)
$\left(\frac{\langle +0 \rangle}{\langle -0 \rangle}\right)$	Ratio of positive zero to negative zero	Positively signed non-distributive zero (PSNDZ)
$\left(\frac{\langle -0 \rangle}{\langle +0 \rangle}\right)$	Ratio of negative zero to positive zero	Negatively signed non-distributive zero (NSNDZ)

Table 5. Nomenclature the basic members of the zeroth number field.

94

Zeroth Resonance Theorem

Formally, the zeroth resonance theorem can be stated as follows.

Let b be a real number. Then

1. *If* $X' = \phi\left(\dfrac{N}{0}\right)$ *where ϕ is an odd number and N is a real number then*

    i. $\log_b 0 = \langle -0 \rangle$ *where $\langle -0 \rangle$ is the negative zero.*

    The result of this equation represents the value of the donor-surrogate set, S_o.

    ii. $\log_{-b} 0 = \langle +0 \rangle$ *where $\langle +0 \rangle$ is the positive zero.*

    The result of this equation represents the value of the interacting surrogate set, S.

2. *On the other hand, if*
    a. $X' = \phi x$
    b. $X' = M \pm m$ *provided $M \neq 0$ where M is a positive integer and $0 < m < 1$.*
    c. $X' = \log_{-b}(-1)$
    then

    i. $\log_{-b}(-1) = \log_b 1 = \langle \tilde{0} \rangle \equiv \langle \tilde{1} \rangle$ ii $\log_{-b} 1 = \log_b -1 = \langle \tilde{0} \rangle \equiv \langle \tilde{1} \rangle$

    where $\langle \tilde{0} \rangle$ is the mean zero, unsigned or distributive zero and $\langle \tilde{1} \rangle$ the neutral unit integer . The result of this equation represents the value of the mean recipient set, $\overline{R}$.

3. *By comparison the*

    i. *Mean zero, unsigned or distributive zero is also defined as* $\langle \tilde{0} \rangle = \langle +0 \rangle + \langle -0 \rangle$

    ii. *'Void zero' is defined as* $\langle \phi \rangle = \langle +0 \rangle - \langle +0 \rangle = \langle -0 \rangle - \langle -0 \rangle = \langle +0 \rangle + \langle -0 \rangle$
    where $\langle \phi \rangle$ is the 'void zero'.

    iii. (a) *Positive non-distributive zero is defined as* $\log_{-b} b = \dfrac{\log_{-b} 0}{\log_b 0} = \dfrac{\langle +0 \rangle}{\langle -0 \rangle}$

    (b) *Negative non-distributive zero is defined as* $\log_b(-b) = \dfrac{\log_b 0}{\log_{-b} 0} = \dfrac{\langle -0 \rangle}{\langle +0 \rangle}$

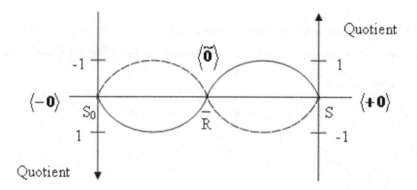

Figure 50 b. Standing set's surrogate waves for the logarithmic quotient of zero. It also represents the logarithm of zero divided by zero.

The illustration of the said conditions under the consideration of the zeroth resonance theorem is shown in figure 50*b* above.

The Resonance Nature of a Zeroth Number Field

The idea that the zeroth number field oscillates has been suggested by numerous equations as was indicated earlier. Here, an attempt will be made to proof this resonance characteristic of the zeroth number field.

By definition, from equation $(26 - 6)$

$$\log_{-10} 0 = \langle + 0 \rangle$$

Let ψ be an arbitrary number. Then

$$\pm \Psi \times 0 = 0$$

provided

$$\pm \Psi \neq 0$$

Thus

$$\log_{-10} 0 = \log_{-10}\left(\pm \Psi \times 0\right)$$
$$= \log_{-10} \pm \Psi + \log_{-10} 0$$
$$= \langle + 0 \rangle$$

This implies

$$\log_{-10} \pm \Psi = \langle + 0 \rangle - \log_{-10} 0$$
$$= \langle + 0 \rangle - \langle + 0 \rangle$$

But for lack of co-existence

$$\langle + 0 \rangle - \langle + 0 \rangle = \langle \phi \rangle$$

while for co-existence

$$\langle + 0 \rangle - \langle + 0 \rangle = \langle + 0 \rangle + \langle - 0 \rangle = \langle \tilde{0} \rangle$$

96

Therefore

$$\log_{-10} \pm \Psi = \left\langle \tilde{0} \right\rangle \equiv \left\langle \phi \right\rangle$$

$$provided \quad \pm \Psi \neq 0.$$

The above result is also true when equation (26 - 1) is used instead of equation (26 - 6). Hence, in general

$$\boxed{\begin{array}{l} \log_b \pm \Psi = \log_{-b} \pm \Psi = \left\langle \tilde{0} \right\rangle \equiv \left\langle \phi \right\rangle \\ provided \quad \pm \Psi \neq 0 \end{array}} \tag{27 - 1}$$

where b and –b is any positive and negative number respectively. Alternatively, equation (27 – 1) can be expressed as

$$\pm \Psi = \pm b^{\left\langle \tilde{0} \right\rangle} \equiv \pm b^{\left\langle \phi \right\rangle} \tag{27 - 2}$$

$$provided \quad \pm \Psi \neq 0$$

Using

$$\pm \Psi = \pm b^{\left\langle \tilde{0} \right\rangle} \quad and \quad \left\langle \tilde{0} \right\rangle = \left\langle -0 \right\rangle + \left\langle +0 \right\rangle$$

we get

$$\pm \Psi = \pm b^{\left\langle -0 \right\rangle + \left\langle +0 \right\rangle}$$

But

$$\left\langle +0 \right\rangle = -\left\langle -0 \right\rangle$$

Therefore

$$\pm \Psi = \frac{\pm b^{\left\langle -0 \right\rangle}}{\pm b^{\left\langle -0 \right\rangle}} = \frac{+ b^{\left\langle -0 \right\rangle}}{+ b^{\left\langle -0 \right\rangle}}$$

which results in

$$\boxed{\begin{array}{c} \pm \Psi = 1 \\ provided \\ \pm \Psi \neq 0 \ and \ \left\langle \tilde{0} \right\rangle = \left\langle \phi \right\rangle. \end{array}}$$

The sense of the LHS and the RHS terms at a glance seems different. However, the $\pm$ sign of the LHS represents the co-existence of a positive and negative numerical sense that can be liken to the positive nucleus and negative electron(s) of a neutral atom. The equivalence of $\left\langle \phi \right\rangle$ and $\left\langle \tilde{0} \right\rangle$ indicates the presence of 'oscillatory' and/or resonance nature of the zeroth number field. This means that both $\left\langle -0 \right\rangle$ and $\left\langle +0 \right\rangle$ appear to interchange positions in the zeroth number field by virtue of their 'oscillations'. The partial dissociation of the zeroth number field is caused by +1 and -1. It can be suggested here that within a '**zeroth number field uncertainty**', $\delta 0$ the real number line is analogous to a single point capacitor charged by opposite point charge queues. Here, the conventional zero serves as a point dielectric. Figure 50*a* above depicts this newly described real number line. On the other hand, the condition that

$$\pm\Psi \neq 0$$

suggests that there exist an induced set of numbers close to the induced zero and beyond. Consequently, the conventional positive and negative numbers are attributes of the said '**numeric induction**'. As such, it can be fairly concluded that there exists a '**neutral number group**' whose resonance nature is represented by the resonance sign, $\{\widetilde{\pm}\}$. Hence, the general equation above can be restated as

$$\{\widetilde{\pm}\}\Psi = 1$$
$$provided$$
$$\{\widetilde{\pm}\}\Psi \neq 0 \ and \ \langle\widetilde{0}\rangle = \langle\phi\rangle. \qquad (27-3a)$$

where $\{\widetilde{\pm}\}\Psi$ represents an infinite set of '**neutral number**'. It possesses a positive and negative numerical sense at the same time. The number 1 at the RHS of equation $(27-3a)$ which is equivalent to those of equations $(26-3a)$ and $(26-5)$ must have the same simultaneous positive and negative number sense. The true nature of the said 'neutral number' must thus satisfy equation $(27-3a)$. This can be ascertain by combining the results of equations $(26-31)$, $(27-2)$ and $(27-3a)$. From these equations, we get

$$\frac{0}{0} = \{\widetilde{\pm}\}\Psi = 1$$
$$provided$$
$$\{\widetilde{\pm}\}\Psi \neq 0 \ and \ \langle\widetilde{0}\rangle = \langle\phi\rangle. \qquad (27-3b)$$

The qualitative implications here is that the numeric senses of the infinite set of 'neutral number' must simultaneously

1. co-exist,
2. annihilate each other via resonance and
3. have a unit sized magnitude.

Such infinite set of numbers can be represented by a set of resonating square root number. In other words,

1.
$$\{\widetilde{\pm}\}\Psi = \{\widetilde{\pm}\}\sqrt{+r}$$

   where +r is a positive number. Here, two possible solutions with opposite number sense exist.

2. The co-existence of the numeric senses has no net '**numeric mono-polarity**', that is positive or negative or net '**numeric di-polarity**', that is ±.

98

3. It achieves a constant unit magnitude through a neutral non-distributive process which can be mathematically expressed as,

$$\frac{0}{0} = \frac{\left|+\sqrt{+r}\right|}{\left|-\sqrt{+r}\right|} = \frac{\left|-\sqrt{+r}\right|}{\left|+\sqrt{+r}\right|} = \left|\{\widetilde{\pm}\}\frac{\sqrt{+r}}{\sqrt{+r}}\right| = \{\widetilde{\pm}1\} = \langle\widetilde{1}\rangle$$

where $\langle\widetilde{1}\rangle$ is the '**neutral unit number**' with numeric senses neutralized but co-existing.

One can therefore generally conclude that,

> *The 'neutral unit number' is an infinite set of neutral self non-distributive numbers of the square root of all positive real numbers whose respective magnitudes is a constant equal to one.*

The important concept with the nature of a 'neutral unit number' is the existing number of groups. In a distributive or sharing reaction expressed mathematically as

$$\frac{x}{y}$$

qualitative implications can be derived by using the following guidelines:

1. Express the given distribution in a '**qualitative distribution sequence form**'. That is,

$$\frac{x}{y} = \left(\frac{1}{y}\right)x$$

2. Divide 1 into y equal groups or parts.
3. Add x number of unit group or part together.

With the above qualitative analysis based distributive guidelines, the bizarre situation of division by zero suddenly becomes sensible. For example, $\frac{2}{0}$ can be analyzed as follows,

Analysis 1: $\frac{2}{0} = \left(\frac{1}{0}\right)2$

Analysis 2: Sequentially,

        a.   ½ implies two existing groups of 1.

        b. $\frac{1}{1}$ implies one existing group of 1.

        c.   It can be concluded that $\frac{1}{0}$ implies zero or no existing group of 1.

    Therefore, our unit group here is a zero group.

Analysis 3: The sum of 2 of the zero groups gives zero.

Invariably, it can be stated generally that

$$\frac{n}{0} = n\{\pm 1\} = n\langle \tilde{1} \rangle = \langle \tilde{1} \rangle$$

$$or$$

$$\sum_{i=1}^{n} \langle \tilde{1} \rangle_i = \langle \tilde{1} \rangle$$

(27 – 4)

where $\langle \tilde{1} \rangle$ represents the '**non-distributive cardinality**' of the 'neutral unit number'

The Quantum Axioms of the Zeroth Number Field

The zeroth number field satisfying a set quantum field axioms are defined below.

1. <u>Quantum Addition</u>

   a.

   $$\langle -0 \rangle + n = n + \langle -0 \rangle = \langle -0 \rangle \qquad \text{(Commutative property)}$$

   Proof

   From the LHS term,

   $$\log_{10} 0 + n = \log_{10} 0 + n \log_{10} 10$$
   $$\textit{This gives from RHS above}$$
   $$RHS = \log_{10} 0 + \log_{10} (10)^n$$
   $$= \log_{10} (10)^n + \log_{10} 0$$
   $$= \log_{10} (10)^n \cdot 0$$
   $$= \log_{10} 0$$
   $$= \langle -0 \rangle$$

b.

$$\langle +0 \rangle + n = n + \langle +0 \rangle = \langle +0 \rangle \qquad \text{(Commutative property)}$$

Proof

The proof here is similar to that of 1(a).

c.

$$\langle \tilde{0} \rangle + n = n + \langle \tilde{0} \rangle = \langle \tilde{0} \rangle \qquad \text{(Commutative property)}$$

Proof

From the LHS term,

$$\langle -0 \rangle + \langle +0 \rangle + n = \log_{10} 0 + \log_{-10} 0 + n\log_{-10} -10$$

$$= \log_{10} 0 + \left\{ \log_{-10} 0 + \log_{-10} [-10]^n \right\}$$

$$= \log_{10} 0 + \log_{-10} 0 \cdot [-10]^n$$

$$= \log_{10} 0 + \log_{-10} 0$$

$$= \langle +0 \rangle + \langle -0 \rangle$$

$$= \underline{\underline{\langle \tilde{0} \rangle}}$$

d.

$$\langle \phi \rangle + n = n + \langle \phi \rangle = \langle \phi \rangle \equiv \langle \tilde{0} \rangle \qquad \text{(Commutative property)}$$

Proof

From the LHS term,

$$\langle -0 \rangle - \langle -0 \rangle + n = \log_{10} 0 - \log_{10} 0 + n\log_{10} 10$$

$$= \log_{10} 0 - \log_{10} 0 \cdot (10)^n$$

$$= \log_{10} 0 - \log_{10} 0$$

$$= \langle -0 \rangle - \langle -0 \rangle$$

$$= \underline{\underline{\langle \phi \rangle}}$$

101

On the other hand, from laws of logarithm

$$\log_{10} 0 - \log_{10} 0 = \log_{10}\left(\frac{0}{0}\right)$$

which by definition gives

$$= \langle \phi \rangle \equiv \langle \widetilde{0} \rangle$$

e.

i. $\left(\dfrac{\langle +0 \rangle}{\langle -0 \rangle}\right) + n = \left(\dfrac{\langle +0 \rangle}{\langle -0 \rangle}\right) + \langle -0 \rangle + \langle \widetilde{1} \rangle$ *provided n is an even number*

ii. $\left(\dfrac{\langle +0 \rangle}{\langle -0 \rangle}\right) + n = \left(\dfrac{\langle +0 \rangle}{\langle -0 \rangle}\right) + \langle +0 \rangle + \langle \widetilde{1} \rangle$ *provided n is an odd number*

(Commutative property)

Proof

From the LHS term of 1 e (i),

$$\log_{-10} 10 + n = \log_{-10} 10 + n \log_{-10} -10$$
$$= \log_{-10} 10 + \log_{-10}(-10)^{n}$$
$$= \log_{-10} 10 \cdot (-10)^{n}$$
$$= \log_{-10} -(-10) \cdot (-10)^{n}$$
$$= \log_{-10} -(-10)^{n+1}$$

which can be further expressed as

$$\log_{-10} -(-10)^{n+1} = \log_{-10}(-1) \cdot (-10)^{n+1}$$
$$= \log_{-10} -1 + (n+1)\log_{-10} -10$$

But $\log_{-10} -10 = 1$.

Therefore,

$$\log_{-10} -(-10)^{n+1} = \left(\log_{-10} -1\right) + n + 1$$
$$= \log_{-10} -1 + n \log_{-10}(-10) + 1$$

102

$$\log_{-10}-(-10)^{n+1} = \log_{-10}-1+\log_{-10}(-10)^{n}+1$$

$$= \left\{ \log_{-10}(-1)^{1}\cdot(-10)^{n}\right\}+1$$

$$= \left\{ \log_{-10}(-1)^{1}\cdot\left[(-1)^{n}\cdot10\right]^{n}\right\}+1$$

$$= \left\{ \log_{-10}(-1)^{1}\cdot(-1)^{n}\cdot10^{n}\right\}+1$$

$$= \left\{ \log_{-10}(-1)^{n+1}\cdot10^{n}\right\}+1$$

$$= n\log_{-10}10+\log_{-10}(-1)^{n+1}+1$$

But by definition, $\log_{-10}-10=\left(\dfrac{\langle+0\rangle}{\langle-0\rangle}\right)$

Therefore

$$n\log_{-10}10+\log_{-10}(-1)^{n+1}+1=n\left(\dfrac{\langle+0\rangle}{\langle-0\rangle}\right)+\log_{-10}(-1)^{n+1}+1$$

But, $n\langle+0\rangle=\langle+0\rangle$ (*multiplication quantum property*).

Therefore, we get

$$\dfrac{n\langle+0\rangle}{\langle-0\rangle}+\log_{-10}(-1)^{n+1}+1=\left(\dfrac{\langle+0\rangle}{\langle-0\rangle}\right)+\log_{-10}(-1)^{n+1}+1$$

From equation (26 − 3), $\log_{-10}-1=\log_{-10}-\langle\tilde{1}\rangle=\langle-0\rangle$

From equation (26 − 5), $\log_{-10}1=\log_{-10}\langle\tilde{1}\rangle=\langle+0\rangle$

Therefore,

$$\log_{-10}(-1)^{n+1}=\langle-0\rangle \quad \text{if } n \text{ is an even number}$$

and

$$\log_{-10}(-1)^{n+1}=\langle+0\rangle \quad \text{if } n \text{ is an odd number.}$$

where 1 *is equivalent to* $\langle\tilde{1}\rangle$.

Observe here that the presence of the 'neutral unit number' in the above result indicates that the surrogate set in a signed non-distributive process is also equivalent to 1 no matter how many distributum is available (see the results in equation (27 – 4). Hence,

$$\left(\frac{\langle+0\rangle}{\langle-0\rangle}\right) + n = \left(\frac{\langle+0\rangle}{\langle-0\rangle}\right) + \langle-0\rangle + \langle-1\rangle + \langle\widetilde{1}\rangle \qquad \textit{if } n \textit{ is an even number.}$$

or

$$\left(\frac{\langle+0\rangle}{\langle-0\rangle}\right) + n = \left(\frac{\langle+0\rangle}{\langle-0\rangle}\right) + \langle+0\rangle + \langle-1\rangle + \langle\widetilde{1}\rangle \qquad \textit{if } n \textit{ is an odd number.}$$

where by 'numeric induction',

$$\langle-0\rangle + \langle\widetilde{1}\rangle = \langle-0\rangle + \langle\widetilde{\pm}1\rangle = -1$$

or

$$\langle-0\rangle + \langle\widetilde{1}\rangle = \langle+0\rangle + \langle\widetilde{\pm}1\rangle = -1$$

(where -1 *and* $+1$ *are the 'primo-numeric senses'.)*

f.

$$\textit{i. } \left(\frac{\langle-0\rangle}{\langle+0\rangle}\right) + n = \left(\frac{\langle-0\rangle}{\langle+0\rangle}\right) + \langle-0\rangle + \langle\widetilde{1}\rangle \qquad \textit{provided n is an even number}$$

$$\textit{ii. } \left(\frac{\langle-0\rangle}{\langle+0\rangle}\right) + n = \left(\frac{\langle-0\rangle}{\langle+0\rangle}\right) + \langle+0\rangle + \langle\widetilde{1}\rangle \qquad \textit{provided n is an odd number}$$

(Commutative property)

Proof

The proof here is similar to that of 1(e).

2. <u>Quantum Subtraction</u>

a.

$$\textit{i. } \langle-0\rangle - n = \langle-0\rangle$$

(Non-commutative property)

$$\textit{ii. } n - \langle-0\rangle = -\langle-0\rangle = \langle+0\rangle$$

From the LHS term of 2 a (i),

$$\log_{10} 0 - n = \log_{10} 0 - n \log_{10} 10$$
$$= \log_{10} 0 - \log_{10} (10)^n$$
$$= \log_{10} \frac{0}{(10)^n}$$
$$= \log_{10} 0$$
$$= \langle -0 \rangle$$

On the other hand, from the LHS term of 2 a (ii)

$$n \log_{10} 10 - \log_{10} 0 = -\log_{10} 0 + n \log_{10} 10$$
$$= -\left[\log_{10} 0 - n \log_{10} 10\right]$$
$$= -\left[\log_{10} 0 - \log_{10} (10)^n\right]$$
$$= -\log_{10} \frac{0}{(10)^n}$$
$$= -\log_{10} 0$$
$$= -\langle -0 \rangle$$
$$= \langle +0 \rangle$$

b.

i. $\langle +0 \rangle - n = \langle +0 \rangle$

(Non - commutative property)

ii. $n - \langle +0 \rangle = -\langle +0 \rangle = \langle -0 \rangle$

Proof

The proof here is similar to that of 2 a.

c.

i. $\langle \tilde{0} \rangle - n = \langle \tilde{0} \rangle$

(Commutative property)

ii. $n - \langle \tilde{0} \rangle = \langle \tilde{0} \rangle$

where $\langle \tilde{0} \rangle = -\langle \tilde{0} \rangle$

From the LHS term of 2 c (i),

$$\left(\langle -0\rangle + \langle +0\rangle\right) - n = \log_{10} 0 + \log_{-10} 0 - n\log_{-10} -10$$

$$= \log_{10} 0 + \log_{-10} 0 - \log_{-10}(-10)^n$$

$$= \log_{10} 0 + \log_{-10} \frac{0}{(-10)^n}$$

$$= \log_{10} 0 + \log_{-10} 0$$

$$= \langle -0\rangle + \langle +0\rangle$$

$$= \langle \tilde{0}\rangle$$

On the other hand, from the LHS term of 2 c (ii),

$$n - \left(\langle -0\rangle + \langle +0\rangle\right) = n\log_{-10} -10 - \left(\log_{10} 0 + \log_{-10} 0\right)$$

$$= \log_{-10}(-10)^n - \log_{10} 0 - \log_{-10} 0$$

$$= -\log_{10} 0 - \log_{-10} 0 + \log_{-10}(-10)^n$$

$$= \log_{10} 0 - \left(\log_{-10} 0 - \log_{-10}(-10)^n\right)$$

$$= -\log_{10} 0 - \log_{-10} 0$$

$$= -\left(\log_{10} 0 + \log_{-10} 0\right)$$

$$= -\left(\langle -0\rangle + \langle +0\rangle\right)$$

$$= -\langle \tilde{0}\rangle$$

But, by definition

$$-\langle +0\rangle = \langle -0\rangle \quad and \quad -\langle -0\rangle = \langle +0\rangle$$

Therefore,

$$-\left(\langle -0\rangle + \langle +0\rangle\right) = -\langle -0\rangle - \langle +0\rangle$$

$$= \langle +0\rangle + \left(-\langle +0\rangle\right)$$

$$= \langle +0\rangle + \langle -0\rangle$$

$$= \langle \tilde{0}\rangle$$

Hence,

$$\langle \tilde{0}\rangle = -\langle \tilde{0}\rangle$$

d.

$$\langle\phi\rangle - n = n - \langle\phi\rangle = \langle\phi\rangle \qquad \textit{(Commutative property)}$$

Proof

From the LHS term,

$$
\begin{aligned}
\left(\langle-0\rangle - \langle-0\rangle\right) - n &= \log_{10} 0 - \log_{10} 0 - n\log_{10} 10 \\
&= \log_{10} 0 - \log_{10} 0 - \log_{10} 10^{n} \\
&= \log_{10} 0 - \left\{ \log_{10} 0 + \log_{10} 10^{n} \right\} \\
&= \log_{10} 0 - \log_{10} 0 \cdot (10)^{n} \\
&= \log_{10} 0 - \log_{10} 0 \\
&= \langle-0\rangle - \langle-0\rangle \\
&= \underline{\underline{\langle\phi\rangle}}
\end{aligned}
$$

On the other hand, from the middle term,

$$
\begin{aligned}
n - \left(\langle-0\rangle - \langle-0\rangle\right) &= n\log_{10} 10 - \left(\log_{10} 0 - \log_{10} 0\right) \\
&= \log_{10} 10^{n} - \log_{10} 0 + \log_{10} 0 \\
&= \log_{10} 10^{n} + \log_{10} 0 - \log_{10} 0 \\
&= \log_{10} (10)^{n} \cdot 0 - \log_{10} 0 \\
&= \log_{10} 0 - \log_{10} \\
&= \langle-0\rangle - \langle-0\rangle \\
&= \underline{\underline{\langle\phi\rangle}}
\end{aligned}
$$

or

$$
\begin{aligned}
\log_{10} 0 - \log_{10} 0 &= \log_{10}\left(\frac{0}{0}\right) \\
&= \underline{\underline{\langle\phi\rangle}}
\end{aligned}
$$

e.

i. $\left(\dfrac{\langle+0\rangle}{\langle-0\rangle}\right) - n = n - \left(\dfrac{\langle+0\rangle}{\langle-0\rangle}\right) = \langle+0\rangle$ *provided n is an even number or* $n = 1$

ii. $\left(\dfrac{\langle+0\rangle}{\langle-0\rangle}\right) - n = n - \left(\dfrac{\langle+0\rangle}{\langle-0\rangle}\right) = \langle-0\rangle$ *provided n is an odd number or* $n \neq 1$.

(Commutative property)

Proof

From the LHS term of 2 e (i),

$$\log_{-10} 10 - n = \log_{-10} 10 - n\log_{-10} -10$$
$$= \log_{-10} 10 - \log_{-10}(-10)^n$$
$$= \log_{-10} 10 - \log_{-10}\left[(-1)\cdot 10^n\right]$$
$$= \log_{-10} 10 - \log_{-10}(-1)^n \cdot 10^n$$
$$= \log_{-10} 10 - \left(\log_{-10}(-1)^n + \log_{-10} 10^n\right)$$
$$= \left(\log_{-10}(10) - \log_{-10} 10^n\right) - \log_{-10}(-1)^n$$
$$= \log_{-10} 10^{1-n} - \log_{-10}(-1)^n$$
$$= (1-n)\log_{-10} 10 - \log_{-10}(-1)^n$$
$$= \log_{-10} 10 - n\log_{-10} 10 - \log_{-10}(-1)^n$$

But,

$$\log_{-10} 10 = \left(\dfrac{\langle+0\rangle}{\langle-0\rangle}\right) \quad and \quad n\log_{-10} 10 = \left(\dfrac{\langle+0\rangle}{\langle-0\rangle}\right)$$

Also,

$$\log_{-10}(-1)^n = \langle-0\rangle \qquad if\ n\ is\ an\ even\ number\ or\ n = 1.$$

and

$$\log_{-10}(-1)^n = \langle+0\rangle \qquad if\ n\ is\ an\ odd\ number\ or\ n \neq 1.$$

Therefore,

$$\left(\frac{\langle+0\rangle}{\langle-0\rangle}\right) - n = \left(\frac{\langle+0\rangle}{\langle-0\rangle}\right) - \left(\frac{\langle+0\rangle}{\langle-0\rangle}\right) - \langle-0\rangle = -\langle-0\rangle = \langle+0\rangle$$

if n is an even number or n = 1.

Or,

$$\left(\frac{\langle+0\rangle}{\langle-0\rangle}\right) - n = \left(\frac{\langle+0\rangle}{\langle-0\rangle}\right) - \left(\frac{\langle+0\rangle}{\langle-0\rangle}\right) - \langle-0\rangle = -\langle-0\rangle = \langle+0\rangle$$

if n is an odd number or n ≠ 1.

The result here makes qualitative sense. Here there exists $\langle+0\rangle$ distributum and $\langle-0\rangle$ recipient in a signed non-distributive condition with an external entity desirous to remove anything from the signed non-distributive neighbourhood. The effect is such that either the distributum member(s) is removed or the recipient member(s) is removed. The relationship between the neutral unit integer and real numbers can be established from 1 (e) and 1 (f) as follows,

    i. $n = \langle-0\rangle + \langle\tilde{1}\rangle$ *provided n is an even number.*

    ii. $n = \langle+0\rangle + \langle\tilde{1}\rangle$ *provided n is an odd number.*

In other words,

    i. $\langle\tilde{1}\rangle = n \pm \langle+0\rangle$

    where an even n invokes a plus sign and an odd n invokes a minus sign.

    ii. $\langle\tilde{1}\rangle = n \mp \langle-0\rangle$

    where an even n invokes a minus sign and an odd n invokes a plus sign.

On the other hand, from the LHS term of 2 e (ii),

$$n - \log_{-10} 10 = n\log_{-10} - 10 - \log_{-10} 10$$

$$= \log_{-10}(-10)^n - \log_{-10} 10$$

$$= \log_{-10}\left[(-1)\cdot 10^n\right] - \log_{-10} 10$$

$$= \log_{-10}(-1)^n \cdot 10^n - \log_{-10} 10$$

$$= \log_{-10}(-1)^n + \log_{-10} 10^n - \log_{-10} 10$$

$$= \log_{-10}(-1)^n + n\log_{-10} 10 - \log_{-10} 10$$

Substitute for the terms on the RHS of the above equation using the following values :

$$\log_{-10} 10 = \left(\frac{\langle + 0\rangle}{\langle - 0\rangle}\right) \quad and \quad n\log_{-10} 10 = \left(\frac{\langle + 0\rangle}{\langle - 0\rangle}\right)$$

Also,

$$\log_{-10}(-1)^n = \langle - 0\rangle \qquad if\ n\ is\ an\ even\ number\ or\ n = 1.$$

and

$$\log_{-10}(-1)^n = \langle + 0\rangle \qquad if\ n\ is\ an\ odd\ number\ or\ n \neq 1.$$

Hence,

$$n - \left(\frac{\langle + 0\rangle}{\langle - 0\rangle}\right) = \langle - 0\rangle \qquad if\ n\ is\ an\ even\ number\ or\ n = 1.$$

or

$$n - \left(\frac{\langle + 0\rangle}{\langle - 0\rangle}\right) = \langle + 0\rangle \qquad if\ n\ is\ an\ odd\ number\ or\ n \neq 1.$$

f.

i. $\left(\dfrac{\langle -0 \rangle}{\langle +0 \rangle}\right) - n = n - \left(\dfrac{\langle -0 \rangle}{\langle +0 \rangle}\right) = \langle +0 \rangle$ *provided n is an even number or n = 1*

ii. $\left(\dfrac{\langle -0 \rangle}{\langle +0 \rangle}\right) - n = n - \left(\dfrac{\langle -0 \rangle}{\langle +0 \rangle}\right) = \langle -0 \rangle$ *provided n is an odd number or n ≠ 1.*

(Commutative property)

Proof

The proof for this is similar to that of 2 e.

3. <u>Quantum Multiplication</u>

    a.

$$\langle -0 \rangle n = \langle -0 \rangle$$

Proof

From the LHS term,

$$(\log_{10} 0)n = n\log_{10} 0 = \log_{10} 0^n = \log_{10} 0 = \underline{\underline{\langle -0 \rangle}}$$

    b.

$$\langle +0 \rangle n = \langle +0 \rangle$$

Proof

From the LHS term,

$$(\log_{-10} 0)n = n\log_{-10} 0 = \log_{-10} 0^n = \log_{-10} 0 = \underline{\underline{\langle +0 \rangle}}$$

c.

$$\langle \widetilde{0} \rangle n = \langle \widetilde{0} \rangle$$

Proof

From the LHS term,

$$
\begin{aligned}
\left(\langle -0 \rangle + \langle +- \rangle \right) n &= \left(\log_{10} 0 + \log_{-10} 0 \right) n \\
&= n \log_{10} 0 + n \log_{-10} 0 \\
&= \log_{10} 0^n + \log_{-10} 0^n \\
&= \log_{10} 0 + \log_{-10} 0 \\
&= \langle -0 \rangle + \langle +0 \rangle \\
&= \underline{\underline{\langle \widetilde{0} \rangle}}
\end{aligned}
$$

d.

$$\langle \phi \rangle n = \langle \phi \rangle$$

Proof

From the LHS term,

$$
\begin{aligned}
\left(\langle -0 \rangle - \langle -0 \rangle \right) n &= \left(\log_{10} 0 - \log_{10} 0 \right) n \\
&= n \log_{10} 0 - n \log_{10} 0 \\
&= \log_{10} 0^n - \log_{10} 0^n \\
&= \log_{10} 0 - \log_{10} 0 \\
&= \langle -0 \rangle - \langle -0 \rangle \\
&= \underline{\underline{\langle \phi \rangle}}
\end{aligned}
$$

112

e.

$$\left(\frac{\langle +0 \rangle}{\langle -0 \rangle}\right)n = \left(\frac{\langle +0 \rangle}{\langle -0 \rangle}\right)$$

Proof

From the LHS term,

$$\frac{\left(n\langle +0 \rangle\right)}{\langle -0 \rangle} = \frac{\langle +0 \rangle}{\langle -0 \rangle}$$

where, by definition

$$n\langle +0 \rangle = \langle +0 \rangle$$

f.

$$\left(\frac{\langle -0 \rangle}{\langle +0 \rangle}\right)n = \left(\frac{\langle -0 \rangle}{\langle +0 \rangle}\right)$$

Proof

The proof here is similar to that of 3 e.

Other Properties

1. <u>Distributivity</u>

$$\langle \phi \rangle(\langle -0 \rangle + \langle +0 \rangle) = \langle \phi \rangle \cdot \langle -0 \rangle + \langle \phi \rangle \cdot \langle +0 \rangle$$

Proof

From the LHS term,

113

$$\langle \phi \rangle \cdot \langle \widetilde{0} \rangle = \langle \phi \rangle \log_{10}\left(\frac{0}{0}\right) = \log_{10}\left(\frac{0}{0}\right)^{\langle \phi \rangle}$$

where by definition,

$$\log_{10}\left(\frac{0}{0}\right) = \langle \phi \rangle \equiv \langle \widetilde{0} \rangle$$

But, from the signed zeroth exponentiation

$$\left(\frac{0}{0}\right)^{\langle \phi \rangle} = \frac{0^{\langle \phi \rangle}}{0^{\langle \phi \rangle}} = \langle \widetilde{1} \rangle$$

Thus,

$LHS = \log_{10}\langle \widetilde{1} \rangle = \langle +0 \rangle$ *(see equation (26 - 5)*

From equation (26 - 31), we get

$$RHS = \left[\log_{10}\left(\frac{0}{0}\right)\right] \cdot \langle -0 \rangle + \langle \phi \rangle \log_{10} 0$$

$$= \log_{10}\left(\frac{0}{0}\right)^{\langle -0 \rangle} + \log_{10} 0^{\langle \phi \rangle}$$

But, $\dfrac{0^{\langle -0 \rangle}}{0^{\langle -0 \rangle}} = \langle \widetilde{1} \rangle$ *and* $0^{\langle \phi \rangle} = \langle \widetilde{1} \rangle$

Therefore,

$RHS = \log_{10}\langle \widetilde{1} \rangle + \log_{10}\langle \widetilde{1} \rangle = 2\log_{10}\langle \widetilde{1} \rangle$

Using equation (26 - 5) we get

$RHS = 2\langle +0 \rangle$

which from quantum multiplication gives,

$RHS = \langle +0 \rangle$

Hence,

$$LHS = RHS = \langle +0 \rangle$$

Conclusively, the zeroth number field undergoes distributivity.

2.   Associativity

$$\left(\langle-0\rangle+\langle+0\rangle\right)+n=\langle-0\rangle+\left(\langle+0\rangle+n\right)$$

Proof

From the LHS term,

Using quantum addition property, we get

$$LHS=\langle\widetilde{0}\rangle+n=\langle\widetilde{0}\rangle \quad and \quad RHS=\langle-0\rangle+\left(\langle+0\rangle+n\right)=\langle-0\rangle+\langle+0\rangle=\langle\widetilde{0}\rangle$$

Therefore,

$$LHS = RHS$$

Conclusively, the zeroth number field undergoes associativity.

3.   Signed Zeroth Exponentiation

$$i.\ 0^{\langle-0\rangle}=-\langle\widetilde{1}\rangle=\langle\widetilde{1}\rangle \quad\quad ii.\ 0^{\langle+0\rangle}=+\langle\widetilde{1}\rangle=\langle\widetilde{1}\rangle$$

$$iii.\ 0^{\langle\phi\rangle}=\langle\widetilde{1}\rangle \quad\quad\quad iv.\ 0^{\langle\widetilde{0}\rangle}=-\langle\widetilde{1}\rangle=\langle\widetilde{1}\rangle$$

Proof

From the earlier proof for $0^0=1$ and later in chapter 5 from the logarithm of a number divided by zero under '*Direct Inductive Solution*', we can state 3 (i) and (ii). For 3 (iii), we have

$$0^{\langle\phi\rangle}=0^{\langle-0\rangle-\langle-0\rangle}=\frac{0^{\langle-0\rangle}}{0^{\langle-0\rangle}}=\frac{-\langle\widetilde{1}\rangle}{-\langle\widetilde{1}\rangle}$$

Therefore,

$$0^{\langle\phi\rangle}=\langle\widetilde{1}\rangle$$

On the other hand, for 3 (iii) we have

$$0^{\langle \tilde{0} \rangle} = 0^{\langle -0 \rangle + \langle +0 \rangle} = 0^{\langle -0 \rangle} \cdot 0^{\langle +0 \rangle} = -\langle \tilde{1} \rangle \cdot \langle \tilde{1} \rangle$$

Therefore,

$$0^{\langle \tilde{0} \rangle} = -\underline{\underline{\langle \tilde{1} \rangle}}$$

4. Inverse 1

$$\langle -0 \rangle + \{-\langle -0 \rangle\} = \{-\langle -0 \rangle\} + \langle -0 \rangle = \langle \tilde{0} \rangle$$

Proof

$$LHS = \langle -0 \rangle + \left(-\langle -0 \rangle\right) = \langle -0 \rangle + \langle +0 \rangle = \langle \tilde{0} \rangle$$

Also,

$$RHS = \left(-\langle -0 \rangle\right) + \langle -0 \rangle = \langle +0 \rangle + \langle -0 \rangle = \langle \tilde{0} \rangle$$

Therefore,   LHS = RHS.

Conclusively, the zero number field thus undergoes inverse operation.

5. Inverse 2

$$\langle +0 \rangle \cdot \langle +0 \rangle^{-1} = \langle +0 \rangle^{-1} \cdot \langle +0 \rangle = 1$$

Proof

From the LHS term,

$$LHS = \langle +0 \rangle \cdot \frac{1}{\langle +0 \rangle} = \frac{\langle +0 \rangle}{\langle +0 \rangle} = 1$$

Also,

$$RHS = \langle +0 \rangle^{-1} \cdot \langle +0 \rangle = \frac{1}{\langle +0 \rangle} \cdot \langle +0 \rangle = \frac{\langle +0 \rangle}{\langle +0 \rangle} = 1$$

Therefore,   LHS = RHS.

Conclusively, the zeroth number field undergoes inverse operation.

116

6. <u>Identity 1</u>

$$\langle \tilde{0} \rangle \cdot 1 = 1 \cdot \langle \tilde{0} \rangle = \langle \tilde{0} \rangle$$

Proof

From quantum multiplication, LHS = RHS = $\langle \tilde{0} \rangle$. Therefore, conclusively the zeroth number field undergoes identity operation.

7. <u>Identity 2</u>

$$\langle \tilde{0} \rangle + 0 = 0 + \langle \tilde{0} \rangle = \langle \tilde{0} \rangle$$

Proof

From quantum addition, LHS = RHS = $\langle \tilde{0} \rangle$. Hence, conclusively the zeroth number field undergoes identity property.

A Qualitative and Quantitative Representation of a Number Line

In general, a number line is quantitatively represented pictorially by a straight line on which every point is assumed to correspond to a real number and every real number to a point. A new qualitative and quantitative representation of a number line will be advanced here to pictorially represent a real number line.

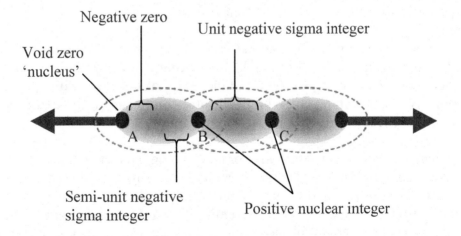

Figure 50 c. Qualitative and qualitative analysis of a number line through nullification of integer nuclei and integer sigma bond into localized integer atoms.

117

Envisaged in the positive section of this newly advanced number line are the positions of all integers will tentatively be represented by queued or chained '**positive nuclear integer**' of '**numeric atoms**'. Found between each pair of 'positive nuclear integers' is what I call a '**negative integer sigma bond**' or '**negative sigma integer**'. The resulting '**numeric sigma bond**' is a consequence of the distributive or sharing interaction that occurs between adjacent 'positive nuclear integers'. Such distributions are therefore dubbed '**sigma distributions**'. From the figure 50*c* above, notice that each unit 'positive nuclear integer' is sandwiched between corresponding semi-unit 'negative sigma integers'. A 'negative sigma integer' is responsible for the micro-divisions between any pair of 'positive nuclear integers'. To understand how this works, consider moving towards the right side of the number line from point A through point B to point C as shown in figure 50*c*. At point A, we have tentatively, a **void zero 'nucleus'** which serves as the nucleus of a zero atom (i.e. a mean zero). To the immediate right of the void zero 'nucleus' is the negative zero followed halfway by a semi-unit 'negative sigma integer'. At point B and C are found 'positive nuclear integers'. Here, the unit 'negative sigma integer' between points A and B cannot balance or nullify the void zero in terms of numeric sense because it has no numeric sense. However, it can nullify the numeric sense of the 'positive nuclear integer' at point B to form what I term a unit '**positive integer atom**' (i.e. a conventional unit positive integer). As one moves from point B to C, a gain in a pair of semi-unit 'negative sigma integer' goes to nullify the positive sense of the 'positive nuclear integer' at point C to another unit 'positive integer atom'. The nullification processes of 'positive nuclear integers' is induced from one 'positive nuclear integer' to another. It is responsible for converting all 'positive nuclear integers' to unit 'positive integer atoms' which together forms the set of '**nullified integers**'. This process is also true for the negative integers of the number. Here, instead of 'positive nuclear integers' and 'negative sigma integers' we have '**negative nuclear integers**' and '**positive sigma integers**' respectively. The result of nullification here are unit '**negative integer atoms**' which together forms the set of '**anti-nullified integers**'.

Integer Hybridization and the Unit Neutral Integer

To understand the nature of the '**neutral unit self non-distributive integer**', one will have to understand the interaction between the first '**negative nuclear integer**' and it's associated '**unit positive type integer**' of the set of '**anti-nullified integers**' and that of the first '**positive nuclear integer**' and its associated '**unit negative type integer**' of the set of '**nullified integers**'.

In a scenario such as a non-distributive or non-sharing interaction, the first 'negative nuclear integer' together with the first 'positive nuclear integer' does not interact with the mean zero 'atom' at all. In other words, no interaction exists between the negative zero of the mean zero 'atom' and it's adjacent '**semi-unit negative type integer**' and also that between the positive zero of the mean zero and its adjacent '**semi-unit positive type integer**'. As a result, the '**unit positive type integers**' around the first 'negative nuclear integer' interact with each other to form a '**unit positive pi integer**' while that of the '**unit**

negative type integers' around the first 'positive nuclear integer' interact with each other to form a '**unit negative pi integer**' as shown in figure 50*d*. It is therefore reasonable to

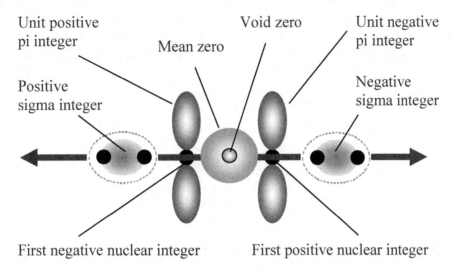

Unit positive pi integer

Mean zero

Void zero

Unit negative pi integer

Positive sigma integer

Negative sigma integer

First negative nuclear integer

First positive nuclear integer

Figure 50 d. Unit positive and unit negative pi integers resulting from a non-distributive process.

conclude that the resulting '**neutral unit integer**' is a representation of what I call an '**anti-numeric bonding**' as a result of a non-distributive interaction between the '**unit positive pi integer**' and the '**unit negative pi integer**' of both first negative and first positive 'nuclear integers' respectively. Such non-distributions are therefore dubbed '**pi distributions**'. The illustration in figure 50*e* below shows the resulting '**numeric pi bonds**' of the said 'neutral unit self non-distributive integer'.

The interaction of unit positive pi integer and unit negative pi integer due to opposite sense attraction to form numeric pi bonds.

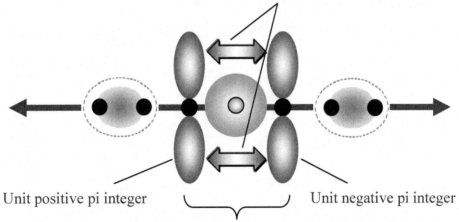

Unit positive pi integer

Unit negative pi integer

The neutral unit self non-distributive integer.

Figure 50 e. The interaction between unit positive pi integer and unit negative pi integer resulting in an anti-numeric bonding due to a non-distributive interaction.

It must be noted here that only the void zero of the zeroth number field exists when the 'neutral unit self non-distributive integer' is formed.

The notion that zero as in the sense of nothingness is based on the concept of singularity as indicated by conventional zero is a misnomer. On the contrary, the internal neighbourhood of conventional zero is based on a plurality concept. Indeed, nothing has something. Figure 50*f* below depicts the place of the zeroth number field on a real number line.

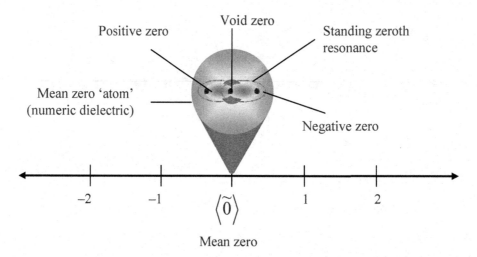

Figure 50 f. Illustration of the concept of zeroth number field on a real number line.

With every point on a real number line assumed to correspond to a real number and every real number to a point, the newly advanced concept of zero which should correspond to a single point suggests that

1. It is really not representative of a true single point.

2. A true single point must exist within a said point on a number line.

A rigorous proof of the above suggestions concerning a geometrical point will be advanced under '*The Continuum Hypothesis*' in chapter 6.

LOGARITHM OF SQUARE ROOT OF NEGATIVE ONE

The trick here is to express the root as a power. Thus

$$\log \sqrt{-1} = \log(-1)^{\frac{1}{2}} = \frac{1}{2}\log(-1)$$

But, by definition

$$\log_b(-1) = \log_{-b}(-1) = \langle\tilde{0}\rangle.$$

Therefore

$$\log \sqrt{-1} = \frac{\langle\tilde{0}\rangle}{2}$$

which is true for any base value.

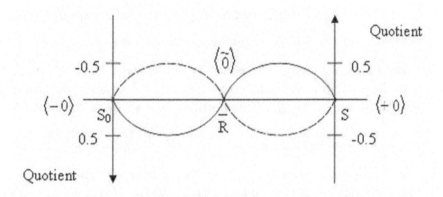

Figure 51. Standing set's surrogate waves for the logarithmic quotient of square-root of -1.

This result which is representative of the quotient means that the quotient's amplitude is reduced from ± 1 to ± 0.5 in relation to the resulting standing wave. Thus, a reduction of resonance takes place here. The graph in figure 51, illustrates all conditions stated.

Observe that the unsigned zero alternates between both positive and negative zero. As such, the solution of the logarithm of square root of negative one which is half of the unsigned zero invokes a polarization effect on the unsigned zero. Rules for the partitioning of the unsigned zero are given below.

The Polarization Rule of Unsigned Zero

The unsigned zero can undergo what is called '**zeroth induction**'. This results in the creation of polarities of zero. The rule governing this process can be generally stated as follows.

121

If the unsigned zero is divided into two equal halves (by dividing by 2), then

1.

$$\frac{\langle \tilde{0} \rangle}{2} = \langle -0 | +0 \rangle$$

where the co-existence of $\langle -0 \rangle$ and $\langle +0 \rangle$ are partially dissociated and $\langle -0 | +0 \rangle$ is called 'zeroth dipole'. This is a partially polarized zero.

2. *The square of the partially polarized zero 'induces' an effectively stronger polarity called 'zeroth dipole pair'. This is a totally polarized zero pair. Thus*

$$\left[\langle -0 | +0 \rangle \right]^2 = \langle \langle -0 \rangle | +0 \rangle \cdot \langle -0 | \langle +0 \rangle \rangle$$

where $\langle -0 \rangle$ and $\langle +0 \rangle$ respectively are called the 'negative zero' and 'positive zero'. They are totally dissociated and represent the effective polarity.

3. *Any number raised to the power of any of the dissociated multiplicands of the totally polarized zero pair, has a corresponding unit magnitude and a sign equal to that of its effective polarity. That is, since by definition $r^0 = 1$*

i. $r^{\langle -0 | \langle +0 \rangle \rangle} = +1$ ii. $r^{\langle \langle -0 \rangle | +0 \rangle} = -1$

where r represents any number. Notice that the expression in 3(i) is equal to that in equation (26-3) and the expression in 3(ii) is equal to that in equation (26-5). Thus,

$$\langle -0 | \langle +0 \rangle \rangle = \langle +0 \rangle \quad and \quad \langle \langle -0 \rangle | +0 \rangle = \langle -0 \rangle$$

Logarithmic Proof of Unsigned Zero Polarization

If, by definition

$$\log_r \sqrt{-1} = \frac{\langle \tilde{0} \rangle}{2}$$

where r is any number and also the base of the logarithm.

Then

$$\sqrt{-1} = r^{\frac{\langle \tilde{0} \rangle}{2}} = r^{\langle -0 | +0 \rangle}$$

Squaring both sides, we get

122

$$-1 = \left[r^{\langle -0|+0\rangle} \right]^2$$

$$-1 = r^{\langle -0|+0\rangle} \cdot r^{\langle -0|+0\rangle}$$

The effect of the totally polarized zero pair, according to the second rule of the polarization rule of the unsigned zero, gives

$$-1 = r^{\langle\langle -0\rangle|+0\rangle} \cdot r^{\langle -0|\langle +0\rangle\rangle}$$

According to the third polarization rule of the unsigned zero

$$r^{\langle\langle -0\rangle|+0\rangle} = -1$$

and

$$r^{\langle -0|\langle +0\rangle\rangle} = +1$$

Hence

$$-1 = r^{\langle\langle -0\rangle|+0\rangle} \cdot r^{\langle -0|\langle +0\rangle\rangle} = (-1)\cdot(+1)$$

which is the result sort.

LOGARITHM OF ZERO DIVIDED BY ZERO

Let,

$$\log\left(\frac{0}{0}\right) = \log 0 - \log 0$$

Then

1. If log is to the base b we have,

$$\log_b\left(\frac{0}{0}\right) = \log_b 0 - \log_b 0$$
$$= \langle -0\rangle - \langle -0\rangle$$
$$= \langle -0\rangle + \langle +0\rangle$$

   But by definition, the presence of both positive and negative zero as a pair represents the 'couple oscillatory flow' and is equivalent to the unsigned zero. Thus,

$$\log_b\left(\frac{0}{0}\right) = \langle \tilde{0}\rangle.$$

2. If log is to the base -b we have,

$$\log_{-b}\left(\frac{0}{0}\right) = \log_{-b} 0 - \log_{-b} 0$$

$$= \langle +0 \rangle - \langle +0 \rangle$$

$$= \langle +0 \rangle + \langle -0 \rangle$$

$$= \langle \tilde{0} \rangle$$

As the results show, the logarithmic quotient value for the log of 0/0 is an unsigned zero. This process removes the ambiguity that hitherto existed. Also, figure 51 represents this situation.

LOGARITHM OF DIVISION BY ZERO

The analysis of a unit division by zero will first be considered here.

1. For the case of a positive unit division by zero, we have the following:

    a. $\log_{b}\left(\dfrac{1}{0}\right) = \log_{b} 0^{-1} = (-1)\cdot\log_{b} 0 = -\langle -0 \rangle = \langle +0 \rangle$

    b. $\log_{-b}\left(\dfrac{1}{0}\right) = \log_{-b} 0^{-1} = (-1)\cdot\log_{-b} 0 = -\langle +0 \rangle = \langle -0 \rangle$

2. For the case of a negative unit division by zero, we have the following:

    a. $\log_{b}\left(\dfrac{-1}{0}\right) = \log_{b}\{(-1)\cdot(0^{-1})\} = \log_{b}(-1) + \log_{b} 0^{-1} = \log_{b}(-1) - \log_{b} 0$
    which gives

    $$\log_{b}\left(\frac{-1}{0}\right) = \langle \tilde{0} \rangle - \langle -0 \rangle = [\langle +0 \rangle + \langle -0 \rangle] - \langle -0 \rangle = \langle +0 \rangle$$

    b. $\log_{-b}\left(\dfrac{-1}{0}\right) = \log_{-b}\{(-1)\cdot(0^{-1})\} = \log_{-b}(-1) + \log_{-b} 0^{-1} = \log_{-b}(-1) - \log_{-b} 0$
    which can be expressed as

    $$\log_{-b}\left(\frac{-1}{0}\right) = \langle \tilde{0} \rangle - \langle +0 \rangle = [\langle +0 \rangle + \langle -0 \rangle] - \langle +0 \rangle = \langle -0 \rangle$$

124

Thus, in general

$$\log_{\pm b}\left(\frac{1}{0}\right) = \log_{\mp b}\left(\frac{-1}{0}\right)$$

The illustration of the above results in cases 1 and 2 can be found in figure 52 below.

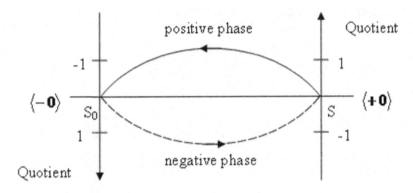

Note: The mean recipient set does not exist in this situation.

Figure 52. Set's surrogate waves for the logarithmic quotient of unit division by zero.

On the other hand, considering the analysis of division by zero of a number n where $1 < n < -1$ and $n \neq 0$ we have the following:

1. For the case of a positive division by zero, we have the following:

a. $\log_b\left(\frac{n}{0}\right) = \log_b\left(n \cdot 0^{-1}\right) = \log_b 0^{-1} + \log_b n$

$$\log_b\left(\frac{n}{0}\right) = \log_b\left(n \cdot 0^{-1}\right) = \log_b n - \langle -0 \rangle \tag{28}$$

b. $\log_{-b}\left(\frac{n}{0}\right) = \log_{-b}\left(n \cdot 0^{-1}\right) = \log_{-b} 0^{-1} + \log_{-b} n = -\langle +0 \rangle + \log_{-b} n$

But

$$\log_{-b} n = \frac{\log_b n}{\log_b - b}$$

where $\log_b - b = \log_b(-1)b = \log_b b + \log_b - 1 = 1 + \langle \tilde{0} \rangle$.

125

Thus,

$$\log_{-b} n = \frac{\log_b n}{1 + \langle \tilde{0} \rangle}$$

Therefore,

$$\log_{-b}\left(\frac{n}{0}\right) = \frac{\log_b n}{1 + \langle \tilde{0} \rangle} - \langle +0 \rangle \qquad (29)$$

2. For the case of a negative division by zero, we have the following:

a. $\log_b\left(\dfrac{-n}{0}\right) = \log_b\left[(-n)\cdot(0^{-1})\right] = \log_b(-n) - \log_b 0$

But

$$\log_b -n = \log_b(-1)n = \log_b n + \log_b(-1)$$

Therefore

$$\log_b\left(\frac{-n}{0}\right) = \log_b n + \log_b(-1) - \log_b 0 = \log_b n + \langle \tilde{0} \rangle - \langle -0 \rangle$$

This is expressed as

$$\log_b\left(\frac{-n}{0}\right) = \log_b n + \left[\langle +0 \rangle + \langle -0 \rangle\right] - \langle -0 \rangle = \log_b n + \langle +0 \rangle$$

which gives

$$\log_b\left(\frac{-n}{0}\right) = \log_b n + \langle +0 \rangle \qquad (30)$$

b. $\log_{-b}\left(\dfrac{-n}{0}\right) = \log_{-b}\left[(-n)\cdot(0^{-1})\right] = \log_{-b}(-n) - \log_{-b} 0$

But

$$\log_{-b} -n = \log_{-b}(-1)n = \log_{-b} n + \log_{-b}(-1)$$

Therefore

$$\log_{-b}\left(\frac{-n}{0}\right) = \log_{-b} n + \log_{-b}(-1) - \log_{-b} 0 = \log_{-b} n + \langle \tilde{0} \rangle - \langle +0 \rangle$$

This implies

126

$$\log_{-b}\left(\frac{-n}{0}\right) = \log_{-b} n + \left[\langle +0 \rangle + \langle -0 \rangle\right] - \langle +0 \rangle = \log_{-b} n + \langle -0 \rangle$$

But by definition

$$\log_{-b} n = \frac{\log_b n}{\log_b(-b)}$$

where, $\log_b -b = \log_b(-1)b = \log_b b + \log_b(-1) = 1 + \log_b(-1)$.

Therefore

$$\log_{-b} n = \frac{\log_b n}{1 + \log_b(-1)}$$

which can be written as

$$\log_{-b}\left(\frac{-n}{0}\right) = \frac{\log_b n}{1 + \log_b(-1)} + \langle -0 \rangle$$

This yields

$$\log_{-b}\left(\frac{-n}{0}\right) = \frac{\log_b n}{1 + \langle \tilde{0} \rangle} + \langle -0 \rangle = \frac{\log_b n}{1 + \langle \tilde{0} \rangle} + \langle -0 \rangle \qquad (31)$$

The denominator in equations (29) and (31), under the negative base analysis, can be expressed respectively as

$$\langle \tilde{0} \rangle + 1 = \langle -0 \rangle - \langle -0 \rangle + 1$$
$$= \langle +0 \rangle - \langle +0 \rangle + 1$$

On a number line, the number 1 will start from $\langle \tilde{0} \rangle$ and end at the mark of 1 on the number line as shown below in figure 53.

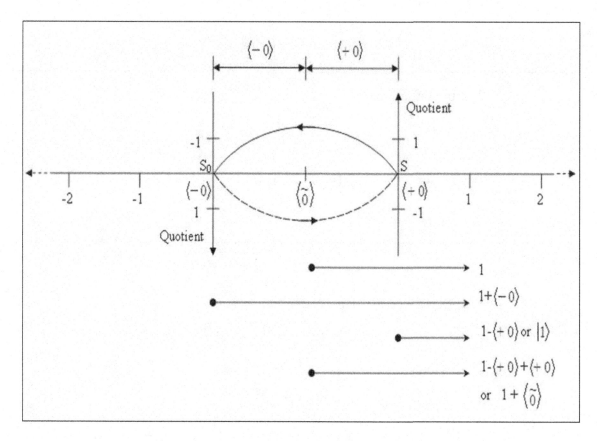

Figure 53. A standing set's surrogate waves for the logarithmic quotient of a general division by zero expressed as a number line.

Observe that the distance $1+\langle -0\rangle$ is equivalent to that of $1-\langle +0\rangle$ since

$$1-\langle +0\rangle =1+\langle -0\rangle .$$

Also, the distance 1 can be expressed as

$$1=|1\rangle +\langle +0\rangle$$

where $|1\rangle$ is the truncated 1 represented by the distance between the set S and the mark 1 on the number line. It is equivalent to that of $1+\langle +0\rangle$ because

$$1+\langle +0\rangle =1-\langle -0\rangle =|1\rangle +\langle +0\rangle -\langle -0\rangle =|1\rangle +\langle \tilde{0}\rangle =1.$$

Since both $\langle -0\rangle$ and $\langle +0\rangle$ represent half the range of $\langle \tilde{0}\rangle$ (see figure 53), adding and subtracting the same kind starting from $\langle \tilde{0}\rangle$ position ends back in $\langle \tilde{0}\rangle$. Thus

$$1+\langle -0\rangle -\langle -0\rangle =1$$

or

$$1+\langle +0\rangle -\langle +0\rangle =1.$$

Consequently, it can be written in terms of interval(s), that

$$\langle -0 \rangle \equiv \langle +0 \rangle.$$

By considering the positive base situations to distinguish between $\log_b\left(\dfrac{n}{0}\right)$ and $\log_b\left(\dfrac{-n}{0}\right)$, equations (28) and (30) are respectively rewritten as

$$\log_b\left(\frac{n}{0}\right) = \log_b n - \langle -0 \rangle = \log_b n + \langle +0 \rangle$$

and

$$\log_b\left(\frac{-n}{0}\right) = \log_b n + \langle +0 \rangle$$

where by definition, $\langle \tilde{0} \rangle + 1 \equiv 1$ and $\langle -0 \rangle = -\langle +0 \rangle$.

Similarly, considering the negative base situations, equations (29) and (31) are respectively rewritten as

$$\log_{-b}\left(\frac{n}{0}\right) = \log_b n - \langle +0 \rangle = \log_b n + \langle -0 \rangle$$

and

$$\log_{-b}\left(\frac{-n}{0}\right) = \log_b n + \langle -0 \rangle$$

where by definition, $\langle \tilde{0} \rangle + 1 \equiv 1$ and $\langle -0 \rangle = -\langle +0 \rangle$.

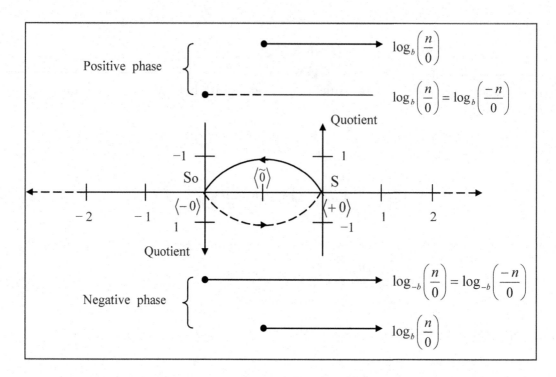

Figure 54. Number line showing distances of various logarithmic quotients under a general division by zero.

The number line in figure 54 illustrates the respective positions of $\log_b\left(\dfrac{n}{0}\right)$ and

$\log_b\left(\dfrac{-n}{0}\right)$ and also $\log_{-b}\left(\dfrac{n}{0}\right)$ and $\log_{-b}\left(\dfrac{-n}{0}\right)$. Observed that since $\left\langle \tilde{0} \right\rangle$ does not exist in

$\log_{\pm b}\left(\dfrac{1}{0}\right)$, it is expected that the values of $\log_b\left(\dfrac{\pm n}{0}\right)$ and $\log_{-b}\left(\dfrac{\pm n}{0}\right)$ would border

around $\left\langle \tilde{0} \right\rangle$. Since

$$\left\langle -0 \right\rangle \equiv \left\langle +0 \right\rangle$$

it can be concluded that

$$\log_b\left(\dfrac{\pm n}{0}\right) \equiv \log_{-b}\left(\dfrac{\pm n}{0}\right).$$

130

Proof of Division by Zero Solution

Earlier in this book, under '*The Question of Preference or Choice*', it was logically deduced that division by zero will realistically result in zero item received by nobody. A rigorous proof will be give below to support the above deduction.

Let x represent a division by zero, then

$$x = \frac{N}{0}$$

where N is a real number.

Expressing the division by zero in a logarithmic quotient form we get

$$\log_b a = x$$

which gives

$$\frac{\log_n a}{\log_n b} = \frac{N}{0}$$

where b, a and n are all real numbers. This means

$$\log_n a = N \text{ which implies } a = N^n$$

and

$$\log_n b = 0 \text{ which implies } b = 1.$$

Thus

$$\log_b a = \log_1 N^n = n \log_1 N \tag{32}$$

But

$$\log_1 N = \frac{\log_{10} N}{\log_{10} 1}$$

This can be generally expressed, since log 1 is always equal to 0, as

$$\log_1 N = \frac{\log_\beta N}{0} \tag{33}$$

where b is a positive real number and $\beta \neq 0$.

Hence, from equations (32) and (33)

$$x = \log_b a = n \log_1 N = \left(\frac{n}{0}\right) \log_\beta N$$

which implies

$$\left(\frac{N}{0}\right) = \left(\frac{n}{0}\right) \log_\beta N$$

This can be written as

$$\left(\frac{N}{0}\right) - \left(\frac{n}{0}\right) \log_\beta N = 0 \tag{34}$$

131

In order to find zero values, N = n.

Hence

$$\left(\frac{N}{0}\right) - \left(\frac{N}{0}\right) \log_\beta N = 0$$

Factorizing

$$\left(\frac{N}{0}\right)\left[1 - \log_\beta N\right] = 0$$

Evaluating the zero values we have firstly

$$1 - \log_\beta N = 0$$

which results in, $N = \beta$. This implies that equation (33) is valid if and only if $N = n = \beta$.

Secondly

$$\left(\frac{N}{0}\right) = 0$$

which is the solution sort.

This result is realistically true. If N items are distributed equally among no existing recipient, it is expected that no recipient gets any item for the result should always be what each recipient gets. As a result, the N items must still be in the hands of the distributor.

Defining the Expressions of 0^0 and 0/0

First of all, the equality between the two expressions can be shown as follows

$$\frac{0}{0} = \frac{0^m}{0^m} = 0^{m-m} = 0^0$$

where m is a real number.

Two solutions will be given to clearly explain the outcome.

1. Indirect Deductive Solution

    The validity of equation (33), is based on the condition that $N \neq 0$ since $N = \beta$ and $\beta \neq 0$. This implies $0/0 \neq 0$ according to equation (33). By definition $x^0 = 1$ for all values of x 'except', under the tentative condition, x = 0. It can be indirectly concluded that when x = 0, 0^0 which is equal to 0/0 must be equal to 1 because $0/0 \neq 0$.

2. Direct Inductive Solution

From equation (34) we can write

$$\left(\frac{n}{0}\right)\log_\beta N = \left(\frac{N}{0}\right)$$

Dividing through by n/0 we get

$$\log_\beta N = \left(\frac{N}{0}\right)\Big/\left(\frac{n}{0}\right)$$

The implication here is that

$$N = \beta^{\left(\frac{N}{0}\right)/\left(\frac{n}{0}\right)} \tag{34}$$

But by definition,

$$N = n = \beta \text{ where } \beta \neq 0.$$

Therefore, for equation (34) to be valid

$$\left(\frac{N}{0}\right)\Big/\left(\frac{n}{0}\right) = 1.$$

This is true since

$$\left(\frac{N}{0}\right) = \left(\frac{n}{0}\right).$$

On the other hand,

$$\left(\frac{N}{0}\right)\Big/\left(\frac{n}{0}\right) = \left(\frac{N}{0}\right)\cdot\left(\frac{0}{n}\right) = \frac{0}{0}.$$

Thus, the implication is that

$$\left(\frac{0}{0}\right) = 0^0 = 1$$

which is the result sort.

The above solution is realistically valid. The reason is that if we let

$$\frac{x}{y} = \frac{0}{0}$$

and the distributive process of x/y is envisaged as how many parts of y are in x then it becomes obvious that there exist one part of 0 in 0 just as there exist one part of 2 in 2. Alternatively, the same result can be proved using logarithm.

The Logarithmic Proof

By definition,

$$\log_b\left(\frac{0}{0}\right) = \langle\tilde{0}\rangle$$

Then

$$\left(\frac{0}{0}\right) = b^{\langle\tilde{0}\rangle}$$

To generalize the base, let b be represented by any number, r. Then it can be stated that

$$r_1^{\langle\tilde{0}\rangle} = r_2^{\langle\tilde{0}\rangle} = r_3^{\langle\tilde{0}\rangle} = \ldots = r_n^{\langle\tilde{0}\rangle} = \left(\frac{0}{0}\right)$$

where 1, 2 and 3 are the first three terms and n is the nth or last term.

The only condition that can satisfy the above equation is when it is equal to 1 since by definition

$$r^0 = 1$$

Hence,

$$r_1^{\langle\tilde{0}\rangle} = r_2^{\langle\tilde{0}\rangle} = r_3^{\langle\tilde{0}\rangle} = \ldots = r_n^{\langle\tilde{0}\rangle} = \left(\frac{0}{0}\right) = 1$$

which is the same result in equation (27-3b). Consequently, to say that there exist 'infinite' parts of 0 in 0 is realistically wrong.

Further Logarithmic Analysis of Division by Zero

If,

$$\log_r\left(\frac{1}{0}\right) = -\langle-0\rangle = \langle+0\rangle$$

where r represents any number.

It implies

$$\left(\frac{1}{0}\right) = r^{\langle+0\rangle}$$

or

$$\left(\frac{1}{0}\right) = r^{-\langle-0\rangle}$$

Using the latter of the two equation of 1/0, it can be stated that

$$\left(\frac{1}{0}\right) = r^{-\langle-0\rangle} = \frac{1}{r^{\langle-0\rangle}}$$

By comparing the LHS to the RHS of the above equation, it is concluded that

$$r^{\langle-0\rangle} = 0$$

Also, if

$$\log_{-r}\left(\frac{1}{0}\right) = -\langle+0\rangle = \langle-0\rangle$$

It implies

$$\left(\frac{1}{0}\right) = -r^{\langle-0\rangle}$$

or

$$\left(\frac{1}{0}\right) = -r^{-\langle+0\rangle} \ .$$

Using the latter of the two equation of 1/0, it can be stated that

$$\left(\frac{1}{0}\right) = -r^{-\langle+0\rangle} = \frac{1}{-r^{\langle+0\rangle}}$$

By comparing the LHS to the RHS of the above equation, it is concluded that

$$-r^{\langle+0\rangle} = 0$$

The results deduced here are also generally true for the case of the logarithm of zero.

LOGARITHM OF ZERO TO THE POWER ZERO

Let,

$$\log\left(0^0\right) = 0\log 0$$

Then, if

1. base of logarithm is positive we get

$$0 \cdot \log_b 0 = 0 \cdot \langle -0 \rangle$$

2. base of logarithm is negative we get

$$0 \cdot \log_{-b} 0 = 0 \cdot \langle +0 \rangle$$

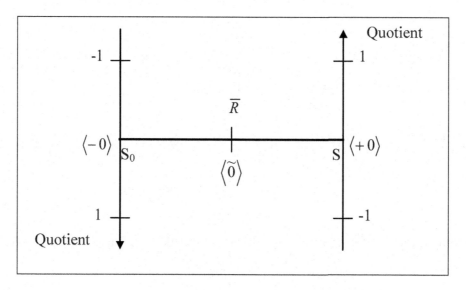

Figure 55. A standing set's surrogate wave beat.

The effect of the results above is what will be dubbed 'set surrogate wave beat'. The zero factor multiplying $\langle +0 \rangle$ and $\langle -0 \rangle$ represents the size of the logarithmic quotient. It eliminates the resonance effect. Figure 55 illustrates the above stated conditions.

LOGARITHM OF THE PRODUCT OF ZERO AND A DIVISION BY ZERO

Let,

$$\log 0 \cdot \left(\frac{1}{0}\right) = \log 0 + \log\left(\frac{1}{0}\right)$$

Then for the following conditions, if

1. base of logarithm is positive we get

$$\log_b 0 \cdot \left(\frac{1}{0}\right) = \log_b 0 + \log_b\left(\frac{1}{0}\right)$$
$$= \langle -0 \rangle + \langle +0 \rangle = \langle \widetilde{0} \rangle.$$

2. base of logarithm is negative we get

$$\log_{-b} 0 \cdot \left(\frac{1}{0}\right) = \log_{-b} 0 + \log_{-b}\left(\frac{1}{0}\right)$$
$$= \langle +0 \rangle + \langle -0 \rangle = \langle \widetilde{0} \rangle.$$

Hence, it can be concluded that

$$\log\left(\frac{0}{0}\right) = \log 0 \cdot \left(\frac{1}{0}\right)$$

which by comparison gives

$$\left(\frac{0}{0}\right) = 0 \cdot \left(\frac{1}{0}\right) = 1$$

because the results in both (1) and (2) are similar to those for logarithm of 0/0.

CHAPTER 6

BIJECTIVE DISTRIBUTION

The formal investigation into the power set analysis of π will be considered through one-to-one correspondence of related power sets.

ANALYSIS OF PI

In geometry, the value of pi is by definition given as

$$\frac{\Omega}{\Theta} = \pi$$

where Ω and Θ are the respective circumference and diameter of a circle and π a constant called pi. Let us consider a simple division process expressed in the usual mathematical symbolism, say

$$\frac{6}{2} = 3$$

Notice that

1. The number 2 represents the members of the recipient set, R (i.e. distribient).

2. The number 6 represents the net members of the distributive power set, $P(\Delta)$ (i.e. dividend or distributum) which is the amount of entities to be equitably shared.

3. The number 3 represents the receivable set, Δ (i.e. quotient or quota) which is the amount of grouped entities each member of the recipient set receives.

Observe that each member of the recipient set, R will have in possession a group with like elements called the receivable set, Δ_x. Hence, each recipient's member is mapped onto a set containing like entities on a one-to-one correspondence. At the conclusion of the equity distribution process, each member of the recipient set will be associated with a set of identical receivables through transfers. This transforms the recipient set into a power set (i.e. a set of sets). The bijection (i.e. one-to-one intersection correspondence) between the recipient set and the receivable set is mathematically expressed as

$$R \cap \Delta_x = \Delta_\varepsilon$$

where Δ_ε is the distributive power set which contains all equally-grouped-distributable entities.

By determining the membership size of each set in the above relation using the distributive ratio of 6/2, the proof of the validity of the relation is given as

$$n(R) \cdot n(\Delta_x) = n(\Delta_1 + \Delta_2 + \Delta_3 + ... + \Delta_x) = n\sum_{k=1}^{x} \Delta_k$$

where k takes positive integer values. Therefore, we get

$$2 \times 3 = 6.$$

Hence, the intersecting relation is valid and it definitively defines the said power set. According to the above intersecting relation, the members of the power set which results due to the intersection between the sets R and Δ_x, is defined equivalently as

$$\{\Delta_1, \Delta_2, \Delta_3, ..., \Delta_x\} = \Delta_\varepsilon$$

where the integer x represents the cardinality of R which is expressed as

$$x = n(R)$$

Also, the relationship between the equivalent member groups or equivalent sets is expressed mathematically as

$$\Delta_1 = \Delta_2 = \Delta_3 = ... = \Delta_x$$

where the integer x represents the cardinality of R which is expressed as:

$$x = n(R)$$

Since the set members of this power set are equal, the power set is called '**equi-power set of distributables**' or simply '**distributive equi-power set**', P (Δ_ε).

Mapping Power Set of Points

Consider the circle shown in figure 56. A one-to-one mapping of mathematical points from the circumference to the diameter shows that every geometrical point on the semi-circumference has a corresponding geometrical point on the diameter. However, the semi-circumference is lengthier than the diameter. Therefore, it can be concluded that the geometrical points on the diameter is denser than that on the arc of the semi-circle.

The mapping of the circumference onto the diameter represents the ratio Ω/Θ. Figure 57 shows a mapping depicting the ratio 6/6. As can be seen, the diameter ba is segmented into 6 equal parts, namely Δ_1, Δ_2, Δ_3, ..., Δ_6. On the other hand, the circumference is divided in 6 equal arcs, namely C_1, C_2, C_3, ..., C_6 which respectively borders 6 sectors, namely 1, 2, 3, ..., 6. The equivalent sets of points in each equal segment on the semi-circumference are

139

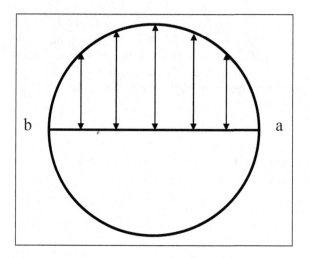

Figure 56. The one-to-one mapping of mathematical points on semi-circumference and its diameter.

mathematically expressed as

$$C_1 = C_2 = C_3 = ... = C_x$$

where the integer x represents the cardinality of R which is expressed as

$$x = n(R)$$

On the other hand, the '**equi-power set of sector arcs**', P (C_ε) on the semi-circumference is defined as

$$\{C_1, C_2, C_3, .., C_x\} = P(C_\varepsilon)$$

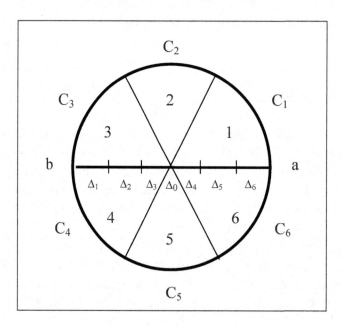

Figure 57.

140

The equivalent sets Δ_1, Δ_2, Δ_3, ..., Δ_6 represents the members of the power set of the recipient(s) while equivalent sets C_1, C_2, C_3, ..., C_6 represents the members of the power set of the distributum. As the radius of the circle is reduced to zero, the segments of the diameter (which are sets) are reduced to a single point, Δ_0 at the center of the circle. Δ_0 represents a power set whose members are all null sets. This power set is dubbed '**null power set**', $P(\Phi_\varepsilon)$. Generally, at Δ_0

$$\Phi_1 = \Phi_2 = \Phi_3 = ... = \Phi_x$$

where the integer x represents the cardinality of R which is expressed as

$$x = n(R).$$

But

$$\{\Phi_1 \cap \Phi_2 \cap \Phi_3 \cap,..,\cap \Phi_x\} = \{\Phi_0\}.$$

Therefore

$$P(\Phi_\varepsilon) = \{\Phi_0\}.$$

The following general summarizations can be made about the radius and diameter of the circle:

> 1. *The points on the circumference of a circle have a one-to-one mapping onto the points of the diameter of the said circumference.*
>
> 2. *The radius has a one-to-one mapping onto a semi-circumference of the said circle when centralized.*
>
> 3. *The centered radius represents an equi-power set.*
>
> 4. *The number of equivalent sets corresponding to the radius must be equal to that of its corresponding semi-circumference.*

Pertaining to the analysis of π, the irretentive equity '**constant distribution**' ratio must be of the following general form,

$$\frac{y}{\psi}$$

where Ψ is a constant equal to 2 called '**di-distribient**' or '**bi-recipient**' and the integer y represents the '**net cardinality**' of the distributive equi-power set, $P(\Delta_\varepsilon)$ which is expressed as

$$y = n\sum_{k=1}^{x} \Delta_k$$

where k takes positive integer values.

141

In the distributive ratio 6/2, two important statements can be made, namely

1. The number 2 which represents the recipients corresponds to the diameter of a circle. This is mathematically expressed as

$$\Theta = 2 \tag{35}$$

2. The number 2 represents an equi-power set with two equal sets. This statement is mathematically expressed as

$$\Theta = 2r \tag{36}$$

where r is the radius of the circle and the 2 at the RHS represents the number of equal sets.

Mapping Members of Choice Sets

In order to maintain an irretentive equity distribution, two crucial things must be done. These are

1. The radius of the circle must be mapped onto a semi-circumference. To achieve this, the radius of the circle is centered as shown in figures 58 and 59.

2. The number of equivalent sets corresponding to the radius must be equal to the number of equivalent sets corresponding to the semi-circumference. Alternatively, the number of equi-power set representing the radius must be equal to the number of equi-power set representative of the semi-circumference.

To map the equivalent sets corresponding to the centered radius, r_c to the corresponding equivalent sets along the semi-circumference, each corresponding common boundary between the two groups of respective equivalent sets is linked with a broken line (for differentiation purposes).

The equi-power set of the centered radius has equivalent sets that are mutually disjoint non-empty sets. Thus, by the stipulation of the *Axiom of Choice*, there exists a choice set that has exactly one member common with each of these sets. [15] In figure 58, the corresponding members of its choice set are $\{\delta_1, \delta_2, \delta_4, \delta_5\}$ while in figure 59 the corresponding members of its choice set are $\{\delta_1$ and $\delta_2\}$. These choice sets represent the common boundaries within the equal segmentation of the centered radius and that of the semi-circumference.

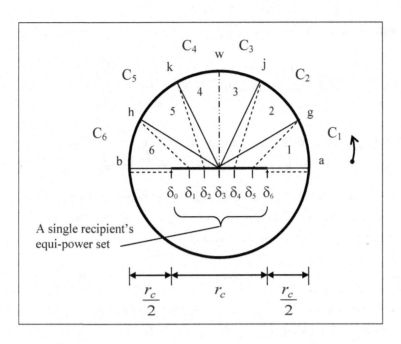

Figure 58. One-to-one equi-power set mapping.

There exist two basic types of mapping linkages. These are,

1. The '**equi-power set radii**', (**EPS** radii). These exist between the equi-power sets of both the centered radius and its corresponding semi-circumference. They are identified by broken lines.

2. The '**null-equi-power set radii**', (**NEPS** radii). These exist between the null power set of the centered radius and the equi-power set of its corresponding semi-circumference. They are identified by continuous lines.

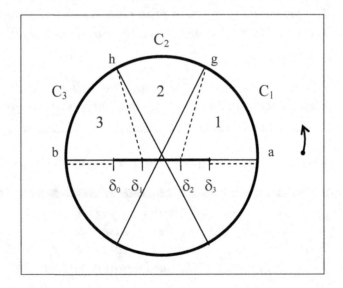

Figure 59.

Mapping Coincidence

This represents the intersection of the two bijections. Similarly in the situation based on earlier discussions, it is the mapping coincidence between the following two one-to-one correspondences,

1. The respective elements of the choice sets of the equi-power sets representing the centered radius, $P(\Delta_\varepsilon)$ and its corresponding semi-circumference, $P(C_\varepsilon)$ and

2. The respective elements of the choice sets of the null power set representing an absolutely diminished centered radius, $P(\Phi_\varepsilon)$ and the equi-power set of its corresponding equi-power set, $P(C_\varepsilon)$.

In the process of general distribution, the recipient(s) get their share or quota only through a form of interaction with the distributor of the entities. This means that every entity received occurred under an intersection process where both recipient(s) and the quota(s) are together. Let us consider a distributive process where,

1. Three items will be distributed to one recipient.

2. The distributor must handout one item at a time to the recipient.

Here, the distributor will have to periodically be together with the recipient in order to deliver the quotas. In other words, the distributor will have to make a one-to-one contact with the recipient three consecutive times. This scenario is called a 'bijective distribution'. In general, two pertinent observations must be remembered, namely

> *1. The periodic intersection of the one-to-one correspondence of a pair of equivalent power sets is indicative of a received quota.*
>
> *2. The periodic count(s) of the unit-by-unit distribution is equal to the number of receivable entities in the quota.*

Observation of the bijective scenarios in both figures 58 and 59 shows that

> *An 'equi-power set radius' coincides periodically with a 'null power set radius' along the anti-clockwise side of the boundary of every third sector, provided more sectors exist else the mapping coincidence occurs once at the said boundary of the third sector.*

Consequently, the general periodicity of the mapping intersection(s) within a '**bi-recipient distribution**' – a general constant distributive situation expressed as ½(y) suggests that,

1. Pi must be a constant and

2. Pi is equal to the '**mapping wavelength**', λ_m which is 3.

The general proof of the above suggestion will be given below.

Since the ratio Ω/Θ is a constant, it implies that the corresponding variation of both Ω and Θ yields the same value of π. For example, by varying the radius in both figures 58 and 59 through decrement, it is evident that a consistent coincidence of the '**mapping line segment**' always occurs at the boundary of every other third sector. In figure 58, two boundary coincidences are found along $\delta_3 w$ and $\delta_0 b$ while in figure 59 it is found along $\delta_0 b$.

Generally,

> *A consistent boundary mapping coincidence occurs when a boundary from the null power set shares the same plane with another boundary from the equi-power set.*

The above generalization will be applied to determine the quota for division processes in the next discussion.

Mapping and the Analysis of Pi Value

Thus far, through mapping frequency and wavelength the mapping coincidence analysis has established a general '**divisional inter-one-to-one relationship**' between the semi-circumference and diameter of a circle.

In the arithmetic of infinity, it is an established fact that

1. There exist an infinite number of numbers or ordinary arithmetical fractions. This is called aleph-null and denoted by $\aleph_0$.

2. The number of all geometrical points on a line is infinite. This is called aleph-one and denoted by $\aleph_1$. It is also referred to as the continuum, c.

3. The infinite number $\aleph_1$ is greater or stronger than the infinite number $\aleph_0$.

It was earlier shown that the number of geometrical points on a semi-circumference is the same as that on the diameter (which is a line) and infinite. This infinite number, which represents the second order of infinite sequence, is denoted by the Hebrew letter $\aleph_1$ (aleph). Here, a general '**geometrical inter-one-to-one relationship**' is seen to exist between the semi-circumference and the diameter of a circle. On the other hand, the number of possible divisional ratios (which are fractions) that can be created using mapping intersections is equal to the infinite number $\aleph_0$.

Let us consider the pairing between the two inter-one-to-one relationships above. It is clear in this '**intra-one-to-one relationship**', that the pairing of the two infinite groups of '**divisional inter-one-to-one relationship**' and '**geometrical inter-one-to-one relationship**' does not yield a one-to-one correspondence since $\aleph_0$ is less than $\aleph_1$. This lack of equality is due to the fact that "the infinity of geometrical points on a line is larger or stronger than the infinity of all integer or fractional numbers" [10] and so the corresponding points between the diameter and the semi-circumference of any given circle

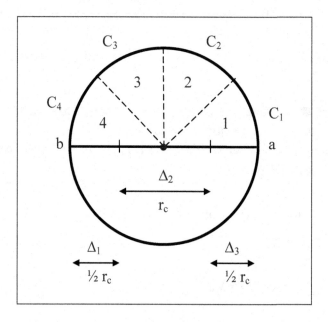

Figure 60.

must be more than all the divisional ratios that can be expressed on the same any given circle. Consequently, for accurate result, the value of the pi ratio (i.e. circumference: diameter) must be based solely on the geometrical points mapping relationship between the semi-circumference and its corresponding diameter.

In figure 60, since it is desired to center the radius of the circle in order to execute the method of mapping intersection, the diameter of the circle is segmented into four equal parts. The two innermost segments represent the centralized radius. For there to be a one-to-one correspondence the semi-circumference is also divided into four equal parts which can either be 4 arcs (represented as C_1, C_2, C_3 and C_4 respectively) or 4 corresponding sectors (represented as 1, 2, 3, and 4 respectively). From infinite arithmetic, the number of geometrical points on lines of different lengths are equal to $\aleph_1$ [11], the geometrical points on the semi-circumference (which is also equal to $\aleph_1$) in figure 60 must have one-to-one correspondence with those on any portion of the diameter ab. However, by definition of the value of pi

146

$$\pi = \frac{\Omega}{\Theta} = \frac{\frac{1}{2}\Omega}{\frac{1}{2}\Theta} = \frac{\frac{1}{2}\Omega}{r_c}$$

where r_c is the radius of a circle.

Therefore, geometrical points on the radius of the circle will be mapped onto corresponding points along the semi-circumference as shown in figure 61. The radius has been centered to enhance the even distribution of geometrical points, arcs and sectors. The centralized radius, in a similar fashion to the diameter, has two equal segments namely, δ_1 and δ_2.

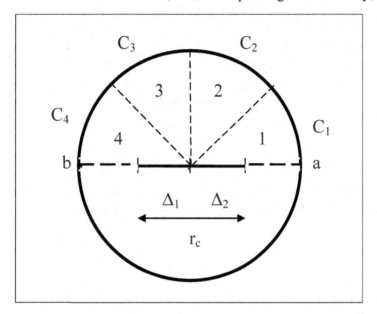

Figure 61.

To determine the '**characteristic mapping relationship**' based on geometrical points, we have the following analysis along the diameter of a circle

1. In figure 61, if 1 radial segment with $\aleph_1$ geometrical points implies 2 equal segments (Δ_1 and Δ_2) each with $\aleph_1$ geometrical points then proportionally
2. In figure 60, 3 radial segments namely D1, D2 and D3 each with $\aleph_1$ geometrical points will imply 6 equal segments each with $\aleph_1$ geometrical points.

By virtue of the desired '**one-to-one characteristic correspondence**', the number of equal sectors or arcs along the semi-circumference must be equal to the effective number of equal segments along the diameter. Thus, the result of 6 equal segments above implies 6 equal sectors or arcs along the semi-circumference corresponding to 6 equal segments along the centralized radius. This newly transformed infinite characteristic segmentation shown in figure 62, establishes a '**characteristic one-to-one relationship**' between the two sets of geometrical points along the circumference and the diameter of a circle.

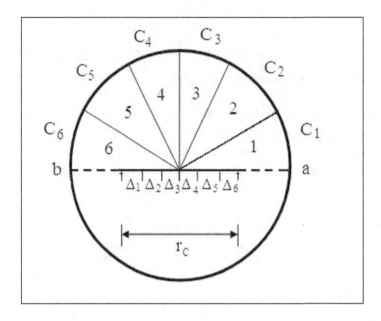

Figure 62.

From figure 62, the mapping intersection can be outlined as was shown earlier on in figure 58 from which the value of pi was computed as 3 using the mapping frequency and wavelength.

MAPPING STATISTICS

The number of consistent boundary mapping coincidence represents '**mapping frequency**', f_b. In figure 58, the mapping statistics (in the upper semi-circle) are as follows

    1. The number of sets of sectors is equal to 6.

    2. The mapping frequency is equal to 2.

    3. The mapping wavelength is equal to 3.

Thus, the equity distributive situation is mathematically expressed as reciprocity of division (i.e. multiplication) by

        Dividend (i.e. distributum) = Recipient (i.e. divisor) x Quota.
which gives
$$6 = 2 \times 3.$$

On the other hand, the mapping statistics (in the upper semi-circle) for figure 59 are as follows

    1. The number of sets of sectors is equal to 3.

148

2. The mapping frequency is equal to 1.

3. The mapping wavelength is equal to 3.

Here too, the equity distributive situation can be mathematically expressed as reciprocity of division by

$$3 = 1 \times 3$$

The equinumerosity (i.e. sameness of cardinality) of the membership sizes can be generalized as follows,

1. *Let*
    i. *$n\Sigma$ be the net cardinality of the equi-power set of distributables, $P(\Delta_e)$.*
    ii. *f_b be the mapping frequency, and*
    iii. *λ_m be the mapping wavelength.*
   Then

   $$\sum n = n\sum\left(P(\Delta_\varepsilon)\right) = n\sum_{k=1}^{x}\Delta_k = f_b \cdot \lambda_m$$

2. *Let*
    i. *$n(P(\Delta_e))$ be the cardinality of the equi-power set of distributables corresponding to the equal segmentation of the centered radius and*
    ii. *$n(P(C_e))$ be the cardinality of the equi-power set of sector arcs corresponding to the equal segmentation of the semi-circumference.*
   Then

   $$n\left(P(\Delta_\varepsilon)\right) = n\left(P(C_\varepsilon)\right)$$

3. Let $n(P(\Phi_\varepsilon))$ be the cardinality of the null power set. Then

   $$n\left(P(\Delta_\varepsilon)\right) = n\left(P(\Phi_\varepsilon)\right)$$

In wave theory, the product of the frequency and the period is always equal to one. The RHS of the first equation of the membership size is equivalent to the product of frequency and the period. It can therefore be proven mathematically that the cardinality of equi-power set is also one. Using equations (35) and (36), it must be noted that

$$2 = 2r$$

which boils down to

$$r = 1$$

where the result 1 represents the cardinality of equi-power set corresponding to the entire centered radius. This power set which has only one member is a '**power singleton**'.

149

Semi-Reduction Principle

In figure 57, the equi-power set corresponding to the diameter is equal to that corresponding to the circumference. At the center of the circle in figure 58, the boundary elements (i.e. members of its choice set) which are actually the '**coinciding members**' of the null power set are mapped correspondingly to the members of the choice set of the equi-power set of the corresponding semi-circumference. However in figure 58, the cardinality (i.e. membership size) of the equi-power set of the centered radius is half that of the diameter, and is equally mapped correspondingly onto those of the equi-power set of the corresponding semi-circumference. Thus, the bijection existing between the null power set at the center of the circle and the corresponding semi-circumference can be mathematically expressed as

$$P(\Phi_\varepsilon) \Rightarrow P(C_\varepsilon)$$

where $P(\Phi_\varepsilon)$ and $P(C_\varepsilon)$ represents the null power set corresponding to the centre of the circle or diameter and the equi-power set corresponding to the semi-circumference.

To achieve one-to-one mapping for the entire circumference, using only the centered radius, a 50% reduction in each segment of the centered radius and its corresponding semi-circumference must take place. This basic equivalent sets' membership down sizing is called '**elemental semi-reduction process**'. Based on the equity ratio distributive scenario in question, that is Ω/Θ, a constant ratio is attained. In terms of cardinality of power set, this is mathematically expressed as

$$\frac{n(C_1)}{n(\Delta_1)} = \frac{n(C_2)}{n(\Delta_2)} = \frac{n(C_3)}{n(\Delta_3)} = ... = \frac{n(C_x)}{n(\Delta_x)} = 1$$

where the integer x represents the cardinality of R.

Consider the paths of the diameter and circumference of a circle as series of geometrical points. Then the following two basic statements can be concluded about a circle

1. The length of its semi-circumference is more than its diameter.

2. The semi-circumference has the same geometrical as its corresponding diameter. Thus, geometrical points on the diameter must be denser.

As a result, the following principle of equi-power set's membership size-reduction can be stated mathematically

$$\frac{\rho_r}{m} = \rho_c$$

which can be expressed in the form of distributive ratio $\Omega/\Theta = \pi$ as

$$\frac{\rho_c}{\rho_r} = \frac{1}{m}$$

where ρ_r and ρ_c are respectively the point densities of the centered radius and its corresponding semi-circumference, and m is a positive integer. It is conclusive from the above comparison that

$$m\pi = 1$$

The multiplier m is referred to here as the '**point-density factor**'.

The reduction principle is illustrated in figure 63 where the centered radius is reduced to 2 segments. It follows from figure 63 that

1. The number of segments on semi-circumference is equal to 3.
2. The mapping frequency is equal to 3 (along $\delta_1 g$, $\delta_1 h$ and $\delta_0 b$).
3. The mapping wavelength is equal to 1.

Thus, from the net cardinality equation of distributive equi-power set

$$n\sum P(\Delta_\varepsilon) = f_b \cdot T_m$$

which gives

$$3 = 3 \times 1.$$

This shows the validity of the reduction principle.

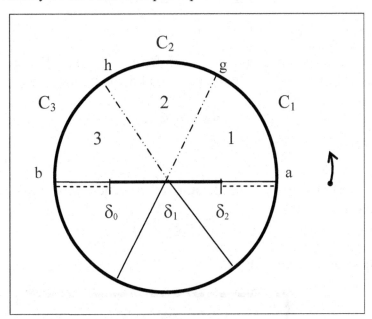

Figure 63.

Also, in figure 64 the segment of the centered radius is increased to 4. Here, too

1. The number of segments on semi-circumference is equal to 3.
2. The mapping frequency is equal to 1½.
3. The mapping wavelength is equal to 2.

151

Notice that the first mapping frequency count is along $\delta_2 h$. This is awarded a mapping frequency of 1/1 because the single coincidence was caused by a single mapping. The other mapping frequency occurs along $\delta_1 b$. But, at b the last mapping also falls along the same path. As a result, it will be awarded a frequency of ½ because the single coincidence was caused by two mappings. This gives a total mapping coincidence frequency of 1½.

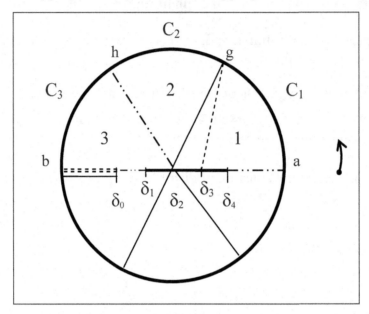

Figure 64.

Alternatively, two equal sectors represent 1 unit frequency value. So, one sector will represent a frequency value of ½. This gives a total mapping frequency value of 1½. Thus, from

$$n(E_s) = f_b \cdot T_m$$

we get

$$3 = 1\frac{1}{2} \times 2$$

which is valid. The constant of proportionality of the general irretentive equity distributive ratio, Ω/Θ varies if

$$n(\Delta_x) \neq n(C_x)$$

In general, if the following conditions

1. $n(\Delta_x) < n(C_x) < n(\Delta_x)$

2. $n(\Delta_x) \Rightarrow n(C_x)$

are true then the mapping is said to be a retentive equity distribution else it is an irretentive equitable distribution.

THE CONTINUUM HYPOTHESIS

Let us consider the contraction of a circle as shown in figure 65. As the radius decreases by a change of a geometrical point at a time, the fraction representing the circumference-diameter ratio (**CDR**) of the infinitely shrinking circle changes in terms of the numerator and the denominator values.

Generally, as the radius shrinks infinitely

1. An infinite series of circumference-diameter ratio is created though each is equal to the same value. This is expressed as: $\dfrac{\Theta_1}{\Omega_1}, \dfrac{\Theta_2}{\Omega_2}, \dfrac{\Theta_3}{\Omega_3}, \dfrac{\Theta_4}{\Omega_4}, ...$

2. The periodic intersection of the circle remains the same. However, the radius of the circle which has $\aleph_1$ geometrical points is infinitely reduced by one geometrical point at a time. The series of geometrical point reduction can be expressed as:
   1, 2, 3, 4, ...

3. There exists a one-to-one correspondence between the infinity of circumference-diameter ratios and the infinity of geometrical points on the radius of the circle to undergo infinite reduction. This isomorphism is shown below.

$$
\begin{array}{cccccccc}
1 & 2 & 3 & 4 & 5 & 6 & 7 & ... \\
\updownarrow & \updownarrow & \updownarrow & \updownarrow & \updownarrow & \updownarrow & \updownarrow & ... \\
\dfrac{\Theta_1}{\Omega_1} & \dfrac{\Theta_2}{\Omega_2} & \dfrac{\Theta_3}{\Omega_3} & \dfrac{\Theta_4}{\Omega_4} & \dfrac{\Theta_5}{\Omega_5} & \dfrac{\Theta_6}{\Omega_6} & \dfrac{\Theta_7}{\Omega_7} & ...
\end{array}
$$

The sequence of geometrical point reduction along the radius of the shrinking radius will take place $\aleph_1$ times. However, the number of fractional circumference to diameter that corresponds to the shrinking radius of a circle is $\aleph_0$. But, $\aleph_0$ is less than $\aleph_1$. Therefore, a natural question that arises is this: how can such varied changes take place simultaneously? It is expected that equal
number of changes must take place simultaneously. To analysis this situation, lets consider figure 65. The number of geometrical points along the reducible radius r_d is $\aleph_1$. Each time the radius gets reduced by a geometrical point, a corresponding new circle is attained. This is illustrated in figure 65 at circle positions $\Theta_2, \Theta_3, ...,$ and Θ_n. A total of $\aleph_1$ different circles are attainable. The key thing to do is to find an answer to the legitimate question: why a total of $\aleph_0$ circumference-diameter fractions are possible out of $\aleph_1$ different circles?

To answer this crucial question, let us examine the point-wise phase changes of the contracting circle. During each phase of a new smaller circle, the net number of geometrical points along its circumference, diameter and area bounded by the circumference will be equal to $\aleph_1$. The limit of decrease in radius of the circle will be a '**single-geometrical-point (SGP) circle**' at the centre Θ_n of the initial circle Θ_1. If '**SGP circle**' is considered as a circle, then the infinity of 'condensed' or 'composite' points in it, dubbed '**aleph-half**' and symbolized as $\aleph_{0.5}$, must be more than the

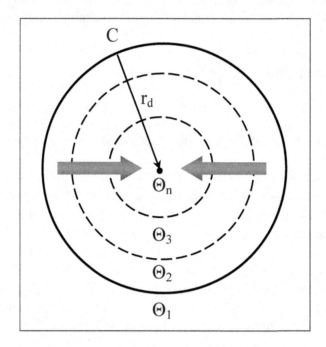

Figure 65. Infinite reduction of the radius of a circle.

number of all integers, $\aleph_0$. Consequently, since $\aleph_1$ is more than $\aleph_0$, there seems reason to suggest that

1. The hidden counts between the circumference-diameter fractions and the corresponding geometrical points along the radius of the circle 'lies' within the center of the circle where the reducing circle will finally rest. As suggested earlier on, this latent counts must be less than $\aleph_0$. Hence,

$$\aleph_{0.5} = \aleph_1 - \aleph_0$$

2. Given that the circumference-diameter ratios (**CDR**) are countable (i.e. its elements can be enumerated), the fact that the infinite cardinal $\aleph_{0.5}$ is hidden strongly suggests that it is uncountable and that must be continuous just like the set of all real numbers.

The transfinite, $\aleph_{0.5}$ must therefore lie between $\aleph_0$ and $\aleph_1$. This result negates the continuum hypothesis which posits that no transfinite number exists between aleph-null and aleph-one. [12]

The latency of a single geometrical point can be generalized as such

A single geometrical point contains infinitely $\aleph_{0.5}$ geometrically latent or ghost points.

Later, the nature of this latency will be elucidated.

Contrary to the infinitely shrinking of a circle, when a circle is infinitely large in size the increasing size of the circle causes the curvature of the circumference of the circle to decrease. Eventually, the limit of the decrease in the curvature will be a straight line. [13] Whereas the infinite contraction process of a circle leads to latently stacked geometrical points, the process of infinite expansion of a circle leads infinitely to the non-stacking of the stacked geometrical points. Here, the latently stacked geometrical points are 'drawn out' or 'distributed' along a real number line.

Formal Definition of a Geometrical Point

The basis of all Geometry is points. Euclid vaguely defined a geometrical point as: *A point is that which has no part, or that which is indivisible.* [14] Three pertinent fundamental statements can be made about a geometrical point, namely

1. Two of the most fundamental concepts in Geometry namely points and lines are the most difficult to define.

2. There is no set definition for them.

3. With the plane, they are the undefined terms of geometry. The concepts of the point, line and plane are the foundations of all other geometric concepts and definitions.

The lack of objects in geometry suggests that a geometrical point is an exact location or position in a plane or space and is dimensionless or zero-dimensional. [9] The above general definition of a geometrical point is better than the vague definition by Euclid. The geometrical point is considered undefined, among other reasons, because of the many equally 'true' definitions.

Three main definitions, among the said equally 'true' definitions, see a point as

1. A dot. As exemplified by a pixel, it dimensional.

2. An exact location. This exemplifies a zero-dimensional point.

3. An ordered pair. These pair numbers help to locate an exact position on a coordinate plane.

155

However, the truthfulness of the set definition for a geometrical point can not be overemphasized, if found. To do this, let us examine the characteristics of the aleph-half and the aleph-one (i.e. the continuum, c).

As explained earlier on, the infinite cardinal $\aleph_{0.5}$ is not countable but continuous like the continuum, c or $\aleph_1$. This explained why it was not accounted for during the infinite contraction of any circle as illustrated in figure 65. The infinite cardinality, $\aleph_{0.5}$ of a geometrical point suggested it is a closely packed or continuously packed concentric locations in space analogous to the continuously packed linear locations of the set of real numbers. Thus,

> ***A geometrical point is a set of itself and is not countable but continuous.***

At first, the above statement seems paradoxical. But in set theory, the fundamental objects of study, which are sets, are determined by the elements that form their respective membership. At the centre of an infinitely contracted circle, the infinite cardinal $\aleph_{0.5}$ that represents the hidden sequence of the point-stages of the infinitely diminishing circle can be put together as an infinite set (or group) shown below.

$$\left\{ \frac{\Theta_{\aleph_0+1}}{\Omega_{\aleph_0+1}}, \frac{\Theta_{\aleph_0+2}}{\Omega_{\aleph_0+2}}, \frac{\Theta_{\aleph_0+3}}{\Omega_{\aleph_0+3}}, \frac{\Theta_{\aleph_0+4}}{\Omega_{\aleph_0+4}}, \frac{\Theta_{\aleph_0+5}}{\Omega_{\aleph_0+5}}, \frac{\Theta_{\aleph_0+6}}{\Omega_{\aleph_0+6}}, \frac{\Theta_{\aleph_0+7}}{\Omega_{\aleph_0+7}}, \ldots \right\}$$

Observed here that since the infinitely countable CDR is aleph-null, the latent-point-wise sequence logically must start from $\aleph_0 + 1$. This must be incremented by one for each geometrical point reduction. Incidentally, there exists a '**continuum equality**' since the value of each of the **CDR** is a constant and equal to pi. This means

$$\pi = \frac{\Theta_{\aleph_0+1}}{\Omega_{\aleph_0+1}} = \frac{\Theta_{\aleph_0+2}}{\Omega_{\aleph_0+2}} = \frac{\Theta_{\aleph_0+3}}{\Omega_{\aleph_0+3}} = \frac{\Theta_{\aleph_0+4}}{\Omega_{\aleph_0+4}} = \frac{\Theta_{\aleph_0+5}}{\Omega_{\aleph_0+5}} = \frac{\Theta_{\aleph_0+6}}{\Omega_{\aleph_0+6}} = \frac{\Theta_{\aleph_0+7}}{\Omega_{\aleph_0+7}} = \ldots$$

Hence, the infinite point set has exactly one member. It is therefore called a '**point singleton**'. This concept negates the idea that a geometrical point has a dimension of zero (i.e. no magnitude or size). [9] Since the cardinality measures size of a set, the cardinality of a '**point singleton**' will be one under continuum equality. This situation satisfies the set definition of a geometrical point. According to the given set definition of a geometrical point, two types of continuum do exist here. These are the '**line continuum**' which describes the set of real number and a '**point continuum**' which describes a geometrical point. As a parting comment, a geometrical point is a point continuum.

The Zeroth Set

On a number line, every geometrical point is represented by a number and every number is represented by a geometrical point. The numbers have both magnitude (i.e. distance from the origin) and direction (i.e. positive or negative). Through a one-to-one correspondence, the number line will be used to help us understand the number zero which is represented by a geometrical point and serves as its origin.

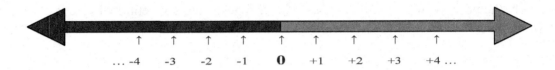

Figure 66 a. A real number line.

Figure 66 a shows a real number line which clearly showing numbers in both directions. To create a one-to-one correspondence, we will match the negative numbers with their corresponding positive numbers. This calls for the bisection of the number line into two equal parts. The resulting mapping is shown below in figure 66 b.

$$
\begin{array}{cccccc}
X_1 & +1 & +2 & +3 & +4 & \dots \\
\updownarrow & \updownarrow & \updownarrow & \updownarrow & \updownarrow & \dots \\
X_2 & -1 & -2 & -3 & -4 & \dots
\end{array}
$$

Figure 66 b. Mapping of positive integers unto corresponding negative integers.

The problem here is to find the value of the points X_1 and X_2. Observe that the result of each positive integer minus 1 is the next positive integer value to the left. Also, it the result of each negative integer plus 1 is the next negative integer value to the left. Therefore,

$$X_1 = +1 - 1 = 0$$

and

$$X_2 = -1 + 1 = 0.$$

Again, it is observed that adding each mapped numbers together yields a zero (i.e. 0). This **'neutralization process of addition'** suggests that the zero resulting is neither negative nor positive. Such a zero was earlier on, under *'Logarithm of Zero'* or *'Zeroth Resonance Theorem'*, called **'mean zero'** or **'unsigned zero'** and shown to be the **'distributional zero'**. For simplicity sake, it is symbolized as $0^{\pm}$ here. Subsequently,

$$X_1 = 0^+$$

and

$$X_2 = 0^-.$$

X_1 and X_2 represent **'positive and negative zeroes'**. These types of zeroes were discussed earlier on under *'Logarithm of Zero'* or *'Zeroth Resonance Theorem'*. The newer symbols

157

used here to represent both positive and negative zeroes are by reason of simplicity. In all, the following '**zeroth arithmetic**'

$$0^+ + 0^- = 0^\pm \leftrightarrow 0$$

can be expressed. The expression above evinces that there exists a one-to-one correspondence between conventional zero (i.e. 0) and the '**mean zero**'. Thus, conventional zero has again been shown to consist of two opposite numeric sense-elements namely the positive and negative zeroes. This suggests that conventional zero is a set with two members. Consequently, it is called the '**zeroth set**'.

Zero-Dimensionality of a Point

Through the veins of set theory, a meaningful physical interpretation explaining how a geometrical point has no dimension will be given.

In a circle, the infinitely of geometrical points that exist is $\aleph_1$. Consequently, the number of diameters that can be drawn is infinitely $\aleph_1$, too. If each diameter is considered as a real number line (which has infinitely $\aleph_1$ geometrical points), then at the centre of the circle, there exist $\aleph_1$ intersections of '**mean zeroes**'. This situation, as depicted in figure 67, is expressed mathematically as

$$\bigcap_1^{\aleph_1} 0^\pm = 0_1^\pm \bigcap 0_2^\pm \bigcap 0_3^\pm \bigcap 0_4^\pm \bigcap 0_5^\pm \bigcap 0_6^\pm \bigcap 0_7^\pm \bigcap \dots$$

Alternatively, it can be expressed as $0 \times 0 \times 0 \times 0 \times 0 \times 0 \times 0 \times \dots$ which is equal to 0. This implies that only one '**zeroth set**' exists at the centre of a circle.

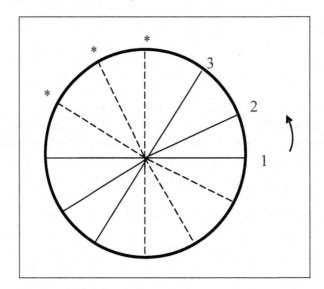

Figure 67. A circle with infinite diameters.

158

As previously shown, the number of elements at the centre of an infinitely diminishing circle is $\aleph_{0.5}$. How does this reconcile with the $\aleph_1$ intersecting 'zeroth sets' at the centre of the circle due to the infinitely $\aleph_1$ diameters? The answer is this. Since both of them exist at the centre of the circle, they must be equivalent. Within the intersections of the 'mean zeroes', the presence of 'zeroth induction' or 'zeroth polarization' (see 'The Polarization Rule of Unsigned Zero') results in the partial dissociation of the 'mean zeroes'. The resulting effect changes the 'mean zero' first to what is dubbed 'zeroth dipoles'. Here, the 'zeroth dipoles' are partially dissociated. Subsequently, through adjacent alignment, the 'zeroth dipoles' become linked thereby creating totally dissociated 'zeroth dipole pairs', $^{+}0^{-}$ as illustrated below in figure 68. Observe that the said linkage is represented as a union between two 'mean zeroes'.

The process of 'zeroth polarization' is responsible for the following

1. The infinite 'zeroth dipole pairs' is less than $\aleph_1$ because an intersection of two 'mean zeroes' is needed for its transformation into 'zeroth dipole pairs'.

2. The lack of one-to-one correspondence between the intersecting members of the infinite 'mean zeroes' set and those of the 'zeroth dipole set' is indicative of the fact that the infinite cardinal representing the two scenarios are not equal.

As it was shown earlier on, the cardinality of a geometrical point at the centre of a circle, due to its infinite shrinkage, is $\aleph_{0.5}$ which is less than $\aleph_1$. Therefore, the cardinality of the 'zeroth dipole' must be $\aleph_{0.5}$. However, the two-to-one correspondence that exists between the 'zeroth dipole set' and the 'zeroth dipole pair' set (see figure 68) couple with the fact that the dipoles originated from the 'zeroth power set' favourably suggest that the 'zeroth dipole set' is uncountably infinite (i.e. continuous). Just as the intersection of the elements of the 'zeroth power set' is equivalent to a 'zeroth power singleton' (i.e. with only one set member), so also are the intersections of the 'zeroth dipole' terms. This elucidates why the infinite $\aleph_{0.5}$ is latent and the nature of a geometrical point at the centre of an infinitely diminishing circle is one of zero-dimension.

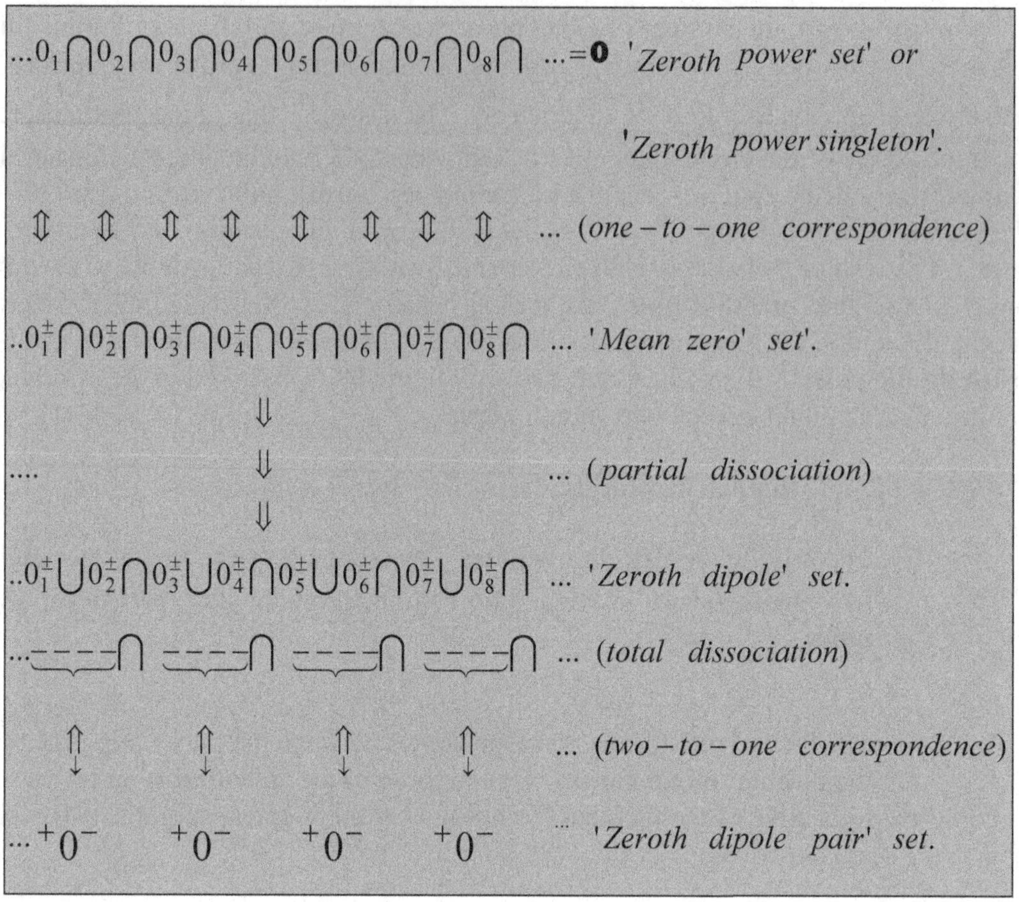

$...0_1 \bigcap 0_2 \bigcap 0_3 \bigcap 0_4 \bigcap 0_5 \bigcap 0_6 \bigcap 0_7 \bigcap 0_8 \bigcap ...= \mathbf{0}$ '*Zeroth power set*' or

'*Zeroth power singleton*'.

$\Updownarrow \quad \Updownarrow \quad \Updownarrow \quad \Updownarrow \quad \Updownarrow \quad \Updownarrow \quad \Updownarrow \quad \Updownarrow \quad ...$ (*one – to – one correspondence*)

$..0_1^{\pm} \bigcap 0_2^{\pm} \bigcap 0_3^{\pm} \bigcap 0_4^{\pm} \bigcap 0_5^{\pm} \bigcap 0_6^{\pm} \bigcap 0_7^{\pm} \bigcap 0_8^{\pm} \bigcap ...$ '*Mean zero*' set'.

$\Downarrow$

.... $\Downarrow$ ... (*partial dissociation*)

$\Downarrow$

$..0_1^{\pm} \bigcup 0_2^{\pm} \bigcap 0_3^{\pm} \bigcup 0_4^{\pm} \bigcap 0_5^{\pm} \bigcup 0_6^{\pm} \bigcap 0_7^{\pm} \bigcup 0_8^{\pm} \bigcap ...$ '*Zeroth dipole*' set.

$...\underbrace{----}\bigcap \underbrace{----}\bigcap \underbrace{----}\bigcap \underbrace{----}\bigcap ...$ (*total dissociation*)

$\Uparrow \qquad\quad \Uparrow \qquad\quad \Uparrow \qquad\quad \Uparrow$ ... (*two – to – one correspondence*)

$...{}^{+}0^{-} \qquad {}^{+}0^{-} \qquad {}^{+}0^{-} \qquad {}^{+}0^{-}$ '*Zeroth dipole pair*' set.

Figure 68. The process of 'zeroth polarization'

CHAPTER 7

FUNDAMENTAL EQUITY DISTRIBUTIONS

Under '*Law of Equity Distribution*', two sets of quotient numbers namely distributive and non-distributive numbers were identified. These describe the quotient of a sharing interaction. While a positive quotient is a distributive number existing in a real number field, a negative quotient is a non-distributive number existing in an abstract number field. Characteristically, a non-distributive number is an indication of the presence of a surrogate activity.

Two basic sets of equity distributions exist. The first basic set consists of

    1.   Irretentive equity distributions and

    2.   Retentive equity distributions.

On the other hand, the second basic set consists of

    1.   Inflationary equity distributions and

    2.   Deflationary equity distributions.

IRRETENTIVE AND RETENTIVE EQUITY DISTRIBUTIONS

So far, the distribution processes analyzed are mostly of irretentive equity distribution. It is a situation where the distributor retains none of the dividends prior to the distribution. On the other hand, the retentive equity distribution is a situation where part or the entire dividend is retained by the distributor prior to the distribution process.

The retained dividend becomes a passive dividend while the remaining dividend which would be equitably distributed becomes the active dividend. This situation can be illustrated using both equations (15) and (16) under '*Inter-Operational Relationships*'. By analyzing the division by zero case, 1/0 using equation (15) we get

$$\underline{\underline{\Pi}}(a \mapsto b) = \left(\frac{a}{Q+1}\right)\underline{\underline{\Sigma}}(a \mapsto b)$$

where $\underline{\underline{\Pi}}(a \mapsto b) = 1 \times 0 = 0$, $\underline{\underline{\Sigma}}(a \mapsto b) = 1 + 0 = 1$, the dividend $a = 1$ and the quotient $Q = \left(\frac{1}{0}\right)$.

This gives

$$Q + 1 = \frac{1 \times 1}{0} = \frac{1}{0} = 0.$$

Thus, $Q = -1$ which can be expressed as, $Q = \left(\frac{1}{0}\right) = -1$.

This result is true under irretentive equity distribution. Here, the negative sign means that the distributor still possesses the distributive item(s) though its retention was not desired before the sharing started. This is a case of '**auto-surrogation**'. It means that once a recipient is identified, the item will be handed over. On the other hand, using equations (16) we have

$$\underline{\underline{\Delta}}(a \mapsto b) = \left(\frac{Q-1}{a}\right)\underline{\underline{\Pi}}(a \mapsto b)$$

where $\underline{\underline{\Pi}}(a \mapsto b) = 1 \times 0 = 0$, $\underline{\underline{\Delta}}(a \mapsto b) = 1 - 0 = 1$, the dividend $a = 1$ and the quotient $Q = \left(\frac{1}{0}\right)$.

This gives

$$Q - 1 = \frac{1 \times 1}{0} = \frac{1}{0} = 0.$$

Thus, $Q = 1$ which is expressed as, $Q = \left(\frac{1}{0}\right) = 1$.

This result is also true for a retentive equity distribution. It means that the distributor prior to distributing the item(s) decided to retain it.

It must be noted that in the earlier proof that gave $N/0 = 0$ where $N \neq 0$, the resulting 0 is representative of what no recipient got. This will be true for either irretentive or retentive equity distribution.

INFLATIONARY AND DEFLATIONARY EQUITY DISTRIBUTIONS

Under the condition of intact item(s) distribution, the ratio x/y represents the distribution of x item(s) among y recipient(s) who will each receive z whole quota of items with w whole item(s) remaining. If the quota of the sharing process is more than what it should normally be in reality, the division is 'inflated' and so the process is called '**inflationary equity distribution**'. On the other hand if the quota of the sharing process is less than what it should normally be in reality, the division is 'deflated' and so the process is called '**deflationary equity distribution**'.

The resulting fractional form of an inflated division is called '**inflated fractional form**' and from the definition of the law of equity distribution under '*The Law of Equity Distribution*' it can be expressed mathematically as

$$\frac{x}{y} = z^+ + \frac{w}{y}$$

where x/y is the primary distributive process and z^+ is the '**inflated quota**'.

For the '**unbiased quota**', z to be less than z^+, either w or y in the secondary distributive process (i.e. w/y) must be negative in order to result in a negative quotient. By definition, a negative quotient is a non-distributive number which exists in an abstract number field. Thus, a surrogate process must be involved. This means that the y found in the secondary distributive process constitute the surrogate set. However, since either w or y in the secondary distributive process can be negative, it is important to ascribe practical meaning to both negative terms. A negative y (from the secondary distributive process) will imply a surrogate distribient while a negative w will imply a surrogate distributum which essentially represents borrowed item(s) from a surrogate distributor. Consequently, the secondary distributive process under inflationary equity distribution is an external distributive process. Since both w and y cannot simultaneously be negative in the secondary distributive process, it calls for the invocation of an '**exclusive surrogation principle**'. It is stated as follows.

> *In a secondary distributive process, the surrogation of both distributum and distribient is mutually exclusive for any given inflationary distributive process.*

The 'inflated fractional form' mathematical expression can thus be rewritten as

$$\frac{x}{y} = z^+ + \frac{w}{y_s} \qquad \text{or} \qquad \frac{x}{y} = z^+ + \frac{w_s}{y}$$

Let the recipient set be R, the distributum or distributor set be D and the net surrogate set be S_T. Then

$$n(R) = y, \; n(D) = x, \; n(S_r) = y_s \text{ and } n(S_d) = w_s$$

where n is the cardinality or number of element(s) in the set(s), S_r is the recipient surrogate set, y_s the amount of secondary recipient(s), S_d the distributum-surrogate set and w_s the amount of secondary distributum. Figure 69 shows a Venn diagram depicting the relationship between the recipient(s) set R, the distributum set D and the net surrogate set S_T under an inflationary condition.

163

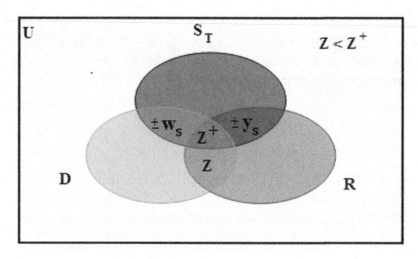

Figure 69.

Here,

$z = n(R \cap D \cap S_T{}')$ or $z = n(R \cap D)$, $Z^+ = n(R \cap D \cap S_T)$, $\pm w_s = n(S_T \cap D)$ and $\pm y_s = n(S_T \cap R)$.

Observe that the possible co-existence between y_s and w_s can be expressed as

$$\frac{w_s}{-y_s} \xrightarrow{\;resonance\;} \xleftarrow[\;equilibrium\;]{} \frac{-w_s}{y_s}$$ (37)

which represents a '**standing set surrogate resonance**' or simply a '**resonance equilibrium**' mode (see '*Concept of Set Resonance*').

On the other hand, the resulting fractional form of a deflated division is called '**deflated fractional form**' and it is expressed mathematically as

$$\frac{x}{y} = z^- + \frac{w}{y}$$

where x/y is the primary distributive process and z^- is the '**deflated quota**'.

For the unbiased quota, z to be greater than z^-, w and y in the secondary distributive process (i.e. w/y) must both be negative or positive in order to give a positive quotient. By definition, a positive quotient is a distributive number which exists in a real number field. This suggests that a surrogate process and a non-surrogate process coexist. The variable *y* found in the secondary distributive process is either a surrogate set (i.e. negative y) or not (i.e. positive y) but cannot be both at the same time. Hence, -*y* is said to be an '**ad hoc surrogate**'. On the other hand, the implication of a positive w is that it is not surrogate in nature while a negative w implies that it is surrogate in nature. A practical description of a negative w in the secondary distributive process is that it is loan given to a surrogate recipient by the distributor. When both positive w and positive *y* coexist, it implies that the

164

process of intact-item distribution facilitated by the distributor was prematurely halted to facilitate leading to a surrogate recipient(s).

A negative *y* (from the secondary distributive process) and a negative w respectively imply a surrogate distribient and a surrogate distributum which essentially represents **loaned item(s)**. Consequently, the secondary distributive process under deflationary equity distribution is also an external distributive process. Since both w and y can be negative at the same time in the secondary distributive process, it calls for the invocation of the **'inclusive net surrogation principle'** which states that

> *In a secondary distributive process, the net surrogation and the net non-surrogation of both distributum and distribient are mutually inclusive for any given deflationary distributive process.*

Here, the 'inflated fractional form' can be rewritten as

$$\frac{x}{y} = z^- + \frac{w_s}{y_s} \qquad \text{or} \qquad \frac{x}{y} = z^- + \frac{w}{y}$$

Letting the recipient set be R, the distributum or distributor set be D and the net surrogate set be S_T we have

$$n(R) = y, \; n(D) = x, \; n(S_r) = y_s \text{ and } n(S_d) = w_s$$

where n is the cardinality or number of element(s) in the set(s), S_r is the recipient surrogate set, y_s the amount of secondary recipient(s), S_d the distributum-surrogate set and w_s the amount of secondary distributum. Figure 70 shows a Venn diagram depicting the relationship between the recipient(s) set R, the distributum set D and the net surrogate set S_T under a deflationary condition.

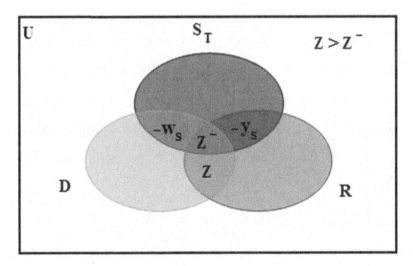

Figure 70.

Here,

$$z = n(R \cap D \cap S') \text{ or } z = n(R \cap D), \; z^{-} = n(R \cap D \cap S), \; -w_s = n(S_T \cap D) \text{ and } -y_s = n(S_T \cap R).$$

Observe here that the possible co-existence between y_s and w_s can be expressed as

$$\frac{-w_s}{-y_s} \xrightarrow{\text{resonance}} \xleftarrow{\text{equilibrium}} \frac{+w_s}{+y_s} \tag{38}$$

Equation (38) represents a '**standing set surrogate resonance**' or simply a '**resonance equilibrium**' mode (see '*Concept of Set Resonance*').

Since the quota is an integer always, an altered quota (z^{+} or z^{-}) must be an integer. The results of inflationary and deflationary division can be outlined based on equation of the law of equity distribution under '*Two Basic Number Fields*'. In general, taking into consideration the sign of x or y we get the following.

For x/y.

1. Inflated Case: $\dfrac{x}{y} = z^{+} + \dfrac{-w}{y}$ where the remainder is negative (e.g. $-w = -6$).

   Example.

   $$\frac{14}{5} = 2 + \frac{4}{5} \equiv 4 \quad + \quad \underbrace{\frac{-6}{5}}_{surrogation}$$

   i.e. $(5 \times 2) + 4 = (5 \times 4) + (-6)$

2. Deflated Case: $\dfrac{x}{y} = \pm z^{-} + \dfrac{w}{y}$ where the remainder is positive (e.g. $+w = +4$).

   Example.

   $$\frac{16}{3} = 5 + \frac{1}{3} \equiv 4 \quad + \quad \underbrace{\frac{4}{3}}_{surrogation} \quad \equiv (-2) \quad + \quad \underbrace{\frac{22}{3}}_{surrogation}$$

   i.e. $(3 \times 5) + 1 = (3 \times 4) + 4 = (3 \times (-2)) + 22$

For x/-y.

3. Inflated Case: $\dfrac{x}{-y} = \pm z^{+} + \dfrac{w}{-y}$

Example.

$$\dfrac{14}{-3} = (-4) + \dfrac{2}{-3} \equiv 2 \quad + \quad \dfrac{20}{\underset{surrogation}{-3}} \quad \equiv (-3) \quad + \quad \dfrac{5}{\underset{surrogation}{-3}}$$

i.e. $\big((-3)\times(-4)\big) + 2 = \big((-3)\times 2\big) + 20 = \big((-3)\times(-3)\big) + 5$

4. Deflated Case: $\dfrac{x}{-y} = -z^{-} - \dfrac{w}{-y}$

Example.

$$\dfrac{3}{-2} = (-2) - \dfrac{1}{-2} \equiv (-5) \quad - \quad \dfrac{7}{\underset{surrogation}{-2}}$$

i.e. $\big((-2)\times(-2)\big) - 1 = \big((-2)\times(-5)\big) - 7$

For -x/y.

5. Inflated Case: $\dfrac{-x}{y} = \pm z^{+} - \dfrac{w}{y}$

Example.

$$\dfrac{-9}{2} = (-4) - \dfrac{1}{2} \equiv -2 \quad - \quad \dfrac{5}{\underset{surrogation}{2}} \quad \equiv 3 \quad - \quad \dfrac{15}{\underset{surrogation}{2}}$$

i.e. $(2\times(-4)) - 1 = (2\times(-2)) - 5 = (2\times 3) - 15$

167

6. Deflated Case: $\dfrac{-x}{y} = -z^- + \dfrac{\pm w}{y}$

Example.

$$\dfrac{-10}{3} = (-3) + \dfrac{-1}{3} \equiv -5 \quad + \quad \underbrace{\dfrac{5}{3}}_{surrogation}$$

i.e. $(3 \times (-3)) + (-1) = (3 \times (-5)) + 5$

Observe that in the deflated case of (6), w is oscillatory in nature.

STABILIZING HYPER AND HYPO-DISTRIBUTIONS

Both inflationary and deflationary distributions need a way to control their inflated and deflated quota respectively.

By definition, a system capable of oscillating such as a string undergoes resonance when an applied force with a frequency equal or nearly equal to one of its natural or characteristic frequencies of the system is set into oscillation with relatively large amplitude. [16] Here, the large amplitude is analogous to a inflated quota. Conversely, if the system is heavily damped, the reduction in amplitude is analogous to a deflated quota.

The phenomenon of set resonance equilibrium is basically analogous to resonance in a string. Characteristics of a resonating system can be generalized as follows:

1. The system must be capable of oscillating. A stretched string has distributional elements each with inertia and elastic characteristics which are analogous to the distributum of a sharing process.

2. A periodic driving force must be applied to keep the system's energy from damping. In a sharing process, such a force comes from the distributor. On the other hand, the distributor can instead cause damping by lending items to a surrogate recipient.

3. The large number of natural frequencies of a string allows resonance to occur at varied frequencies and so is the case of sharing.

4. Energy is absorbed from the vibrational resonance. The natural frequency of the absorbed energy is analogous to the number of surrogate item(s) absorbed via surrogate borrowing.

With the above resonance framework in mind, the following principle is set forth for the determination of the natural frequency of the set resonance equilibrium. By definition, the '**principle of resonance equilibrium**' can be put forth as follows.

Generally, the 'principle of resonance equilibrium' *states that,*

1. *The natural frequency of set resonance equilibrium must result from an interaction between surrogate and non-surrogate sets.*

2. *The relationships between the non-surrogate and surrogate sets must be distributive in nature. This is achieved by using the distributive property of arithmetic operation.*

Inflationary Analysis

Basically, under inflationary condition, for there to be a '**borrowing quotient balance**'

$$z = z^+ \ominus \frac{w}{y} \tag{39}$$

where z is the unbiased quota and $\ominus$ is the negative balancing sign.

Since a larger value in need from w/y to balance the excessive value of z^+, the condition required is $w > y$. The negative nature of the term w/y makes w a borrowed distributum from a surrogate distributor. Thus, in general

---

The distributum surrogate set and the recipient surrogate set are in resonance.

Using equation (37), the natural frequency, υ_{n+} of the inflationary distribution can be expressed as

$$\upsilon_{n+} = \left|\frac{w}{y}\right| = \frac{w_s}{-y_s} \xrightarrow[\text{equilibrium}]{\text{resonance}} \frac{-w_s}{y_s}$$

169

where $\left|\dfrac{w}{y}\right|$ is the natural or characteristic frequency of the inflationary distribution.

From the first principle of resonance equilibrium, the above equation is expressed as

$$v_{n+} = n\left(D_s \cap R\right) \underset{\underset{equilibrium}{\longleftarrow}}{\overset{\overset{resonance}{\longrightarrow}}{\rule{3cm}{0pt}}} n\left(D \cap R_s\right)$$

Applying the second principle of resonance equilibrium, we can write

$$v_{n+} = n\left(D \cap R\right)\left[n\left(D_s\right) + n\left(R_s\right)\right]$$

Expanding gives

$$v_{n+} = n\left(D \cap R\right) \cdot n\left(D_s\right) + n\left(D \cap R\right) \cdot n\left(R_s\right)$$

which can finally be expressed as

$$v_{n+} = \left[n\left(D \cap R\right) \cap n\left(D_s\right)\right] \cup \left[n\left(D \cap R\right) \cap n\left(R_s\right)\right] \qquad (40)$$

Equation (40) is the equation for the inflationary resonance equilibrium's natural frequency. To find the normalized value of the inflated quota z^+, equation (39) in invoked. From equation (39), the equation of normalized quota, z_o^+ is given by

$$z_o^+ = z \oplus \left|\dfrac{w}{y}\right| = z \oplus v_n \qquad (41)$$

where z_o^+ is the normalized inflated quota.

A Venn diagram of four sets using congruent circles does not account for all the needed regions. By using neither congruent, nor convex shapes nor no rotational symmetry, John Venn was able to produce a simple, non-symmetric Venn diagram for four sets. [17] Use was made of congruent and convex ellipses. As an auxiliary illustration, such a diagram will be used to depict the regions of –w and y interaction, w and –y interaction and D and R interaction as shown in figure 71 below.

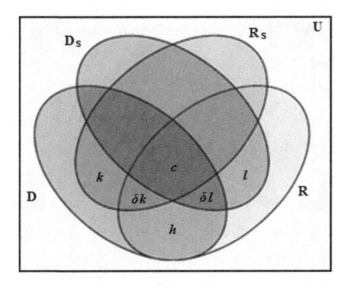

Figure 71. A non-symmetric Venn diagram depicting an inflationary distribution.

Here,

$h = z = n\ (R \cap D)$, $c = z^+ = n\ (R \cap D \cap R_s \cap D_s)$, $l = -w/y = n(R \cap D_s)$ and $k = w/-y = n(R_s \cap D)$.

However, for the convenience of simplicity, the sets of D, R, D_s and R_s are shown in a Venn diagram of four sets using congruent circles in figure 72. The essence here is to be able to adequately illustrate the resonance equilibrium effect and its possible interaction with the quota region. Such a dynamic interaction is of paramount importance. In figure 72,

$h = z = n\ (R \cap D)$, $c = z^+ = n\ (R \cap D \cap R_s \cap D_s)$, $g = -w/y = n(R \cap D_s)$ and $f = w/-y = n(R_s \cap D)$.

It is found that the combined regions of f, a, c, and d analogously defines a loop of partial mode of 'vibration' in relation to the set resonance equilibrium. On the other hand, its counterpart loop is defined by the combined regions of g, b, e and c. The region defined by a, b, c, d and e is firstly responsible for the bonding of the double loops of the set resonance equilibrium region. Secondly, it depicts the mutual interaction between its double loops and the rest of the distributions via the nuclear region c which comprises of all the sets at play. Hence, the double-loop region imitates the resonance wave energy of a string vibrating with its first overtone. The equilibrium state of the double-looped region defined by f, a, c, d, g, b, and e comparably mimics a state of resonance where a system vibrates with maximum energy. Thus, in accordance with the basic characteristics of a resonating system, 'energy' can be absorbed by the region h which borders it.

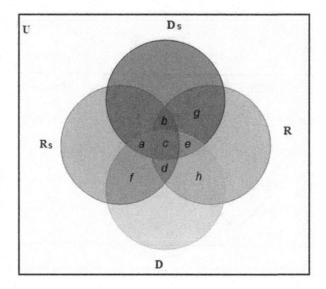

Figure 72. A symmetric Venn diagram for an inflationary distribution.

Now using the areas from figure 72 as equivalent analysis data in equation (41), we get

$$z_o^+ = h \oplus e + d = h + e + d$$

This is the equation for the normalized inflated quota, z_o^+. Ideally, for a borrowing distributional activity to be harmonized and kept under control,

$$z_o^+ < z^+$$

Notice how regions e and d are absorbed from the equilibrium state of the double-looped region which resonates with maximum 'energy' to add up to the unbiased quota.

Deflationary Analysis

Under deflationary condition, for there to be an '**lending quotient balance**' under deflationary condition,

$$z = z^- \oplus \frac{w}{y} \qquad (42)$$

where z is the unbiased quota and (+) is the positive balancing sign.

Since a larger value in need from w/y to balance the shortage of the value of z⁻, the condition required is w > y. The positive nature of the term w/y makes w a lending distributum to a surrogate recipient(s).

In general,

Surrogation and non-surrogation of the secondary distributive process are in resonance.

Using equation (38), the natural frequency, υ_{n-} of the inflationary distribution can be expressed as

$$\nu_{n-} = \left|\frac{w}{y}\right| = \frac{-w_s}{-y_s} \xrightarrow[\underset{equilibrium}{\longleftarrow}]{resonance} \frac{+w_s}{+y_s}$$

where $\left|\dfrac{w}{y}\right|$ is the natural or characteristic frequency of the deflationary distribution.

By the first principle of resonance equilibrium, the above equation is written as

$$\nu_{n-} = n(D_s \cap R_s) \xrightarrow[\underset{equilibrium}{\longleftarrow}]{resonance} n(D \cap R)$$

Applying the second principle of resonance equilibrium, we can write

$$\nu_{n-} = n(D \cap R)[n(D_s) + n(R_s)]$$

Expanding gives

$$\nu_{n-} = n(D \cap R) \cdot n(D_s) + n(D \cap R) \cdot n(R_s)$$

which can finally be expressed as

$$\nu_{n-} = [n(D \cap R) \cap n(D_s)] \cup [n(D \cap R) \cap n(R_s)] \tag{43}$$

Equation (43) is the equation for the deflationary resonance equilibrium's natural frequency. Observe that the natural frequencies of both inflationary and deflationary distributions are equal. Thus

$$\nu_{n+} = \nu_{n-} \tag{44}$$

Using equation (42) to find the normalized value of the deflated quota z ˉ, the normalized deflated quota, z_o^- is given by

$$z_o^- = z \ominus \left|\frac{w}{y}\right| = z \ominus \nu_n \tag{45}$$

where z_o^- is the normalized deflated quota.

The regions of –w and –y interaction, w and y interaction and D and R interaction are shown in a non-symmetric Venn diagram in figure 73 below.

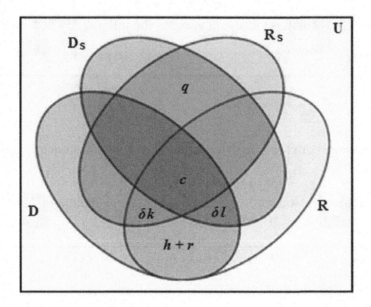

Figure 73. A non-symmetric Venn diagram depicting a deflationary distribution.

Here,

$h = z = n\,(R \cap D)$, $c = z^{-} = n\,(R \cap D \cap R_s \cap D_s)$, $q = -w/-y = n(R_s \cap D_s)$ and $r = w/y = n(R \cap D)$.

Also, the sets of D, R, D_s and R_s are shown in a Venn diagram of four sets using congruent circles in figure 74 to facilitate the illustration of resonance equilibrium effect and its possible interaction with the quota region. In figure 74,

$h = z = n\,(R \cap D)$, $c = z^{-} = n\,(R \cap D \cap R_s \cap D_s)$, $u = -w/-y = n(R_s \cap D_s)$ and $t = w/y = n(R \cap D)$.

Notice that the combined regions of u, a, c, and b analogously defines a loop of partial mode of 'vibration' in relation to the set resonance equilibrium. On the other hand, the alternate loop by virtue of set operative law of idempotent degenerates due to effective damping. The damping effect can be expressed as

$$h = t \;\Rightarrow\; (D \cap R) \cup (D \cap R) = (D \cap R) \quad (\textit{Idempotent Law})$$

is defined by the combined regions of g, b, e and c. The region defined by a, b, c, d and e is firstly responsible for the bonding of the double loops of the set resonance equilibrium region. On the contrary, the existence of mutual interaction between the damped resonance equilibrium and the rest of the distributions is facilitated through the nuclear region c which comprises of all the sets at play. Thus, in accordance with the basic characteristics of a resonating system, a periodic supply of 'energy' is needed to maintain resonance. This must come from the region of D and R interaction.

174

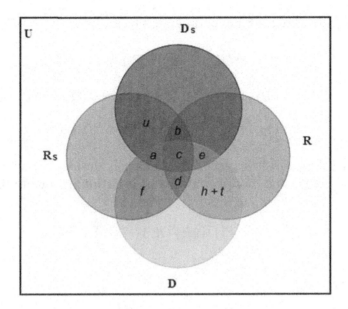

Figure 74. A symmetric Venn diagram for a deflationary distribution.

Now using the areas from figure 74 as equivalent analysis data in equation (44), we get

$$z_o^- = h \ominus e - d = h - e - d$$

This is the equation for the normalized deflated quota, z_o^-. Ideally, for a lending distributional activity to be harmonized and kept under control,

$$z_o^- > z^-$$

Notice how regions e and d relocate absorbed 'energy' from the idempotent quota of region h to revive the damped resonating equilibrium state.

Probability of harmonizing Inflated and Deflated Quota

In figure 71,

$$d = \delta k \text{ and } e = \delta l$$

where δk and δl are the differential k and differential l respectively. It is clear here as to how the union of k, δk, δl and l forms a double-loop region imitative of resonance wave energy of a vibrating string in its first overtone. Since the Venn diagram in figure 71 is accurate, one can calculate the probability of harmonizing an inflated quota, z^+. Thus, from figure 71

$$\delta k = n(D \cap R \cap R_s).$$

As such the probability, P(δk) of the modification factor δk can be written, using De Morgan's law, as

$$P(\delta\ k)= P(D\cap R\cap R_s)= P(D)\cdot P(R)\cdot P(R_s)$$

Similarly, the probability, P(δl) of the modification factor δl can be expressed as

$$P(\delta\ l)= P(D\cap R\cap D_s)= P(D)\cdot P(R)\cdot P(D_s)$$

However, both modifications take place simultaneously during normalization of an unbiased quota, z. Hence, the '**net probability of inflationary normalization**', ΣP_{n+} is given by

$$\bigcap_{n=\delta l}^{\delta k} P_{n+} = P(D\cap R\cap R_s)\cdot P(D\cap R\cap D_s)$$

which evaluates as

$$\bigcap_{n=\delta l}^{\delta k} P_{n+} = 2P(D)\cdot 2P(R)\cdot P(R_s)\cdot P(D_s)$$

On the other hand, from equation (44) the calculation of the harmonizing probability for the deflated quota, z^- is exactly the same as that for the harmonizing probability for inflated quota, z^+. Hence, the '**net probability of deflationary normalization**', ΣP_{n-} is given by

$$\bigcap_{n=\delta l}^{\delta k} P_{n-} = P(D\cap R\cap R_s)\cdot P(D\cap R\cap D_s)$$

which evaluates as

$$\bigcap_{n=\delta l}^{\delta k} P_{n-} = 2P(D)\cdot 2P(R)\cdot P(R_s)\cdot P(D_s).$$

Thus, generally

$$\bigcap_{n=\delta l}^{\delta k} P_{n+} = \bigcap_{n=\delta l}^{\delta k} P_{n-}$$

is the equation representing the general equality between the net inflationary and the net deflationary probability of normalized distribution.

Practical Application

In business, the marketing mix (i.e. product, pricing, place and promotion) has one very important basic component called the distribution channel or place. It is the chain of intermediaries through which the product passes down from one organization to another until it reaches the consumer or end-user.

Acting as the circulatory system of the marketing mix to facilitate the flow of commodity, the monitoring and managing of the distribution channel is very paramount. Generally, the cost of using intermediaries to achieve wider distribution is supposedly lower. However, if the manufacturer (and/or supplier) has any aspirations to be market-oriented, then they should take serious charge in managing all the processes involved in distribution channel until the product or service arrives with the end-user.

The basic types of distribution channels (physical distribution) are,

1. Manufacturer.
2. Distributor - sells to retailers.
3. Retailer - sells to consumers or end-users.

In general, the occurrence of a business distribution channel can be classified as locally (i.e. internal) or foreign (i.e. external). The entire business distribution channel can be broken down into three phases namely '**primary distribution phase**', '**secondary distribution phase**' and the '**tertiary distribution phase**'. The groups involved in each phase are shown in table 6 below.

DISTRIBUTION PHASE	INTERACTING GROUPS	SCOPE DESCRIPTION	SURROGATE GROUP
Primary	Manufacturer and distributor.	Macro-distributor and macro-end-user respectively.	Retailers
Secondary	Distributor and retailers.	Meso-distributor and meso-end-user respectively.	Customers
Tertiary	Retailer and customers.	Micro-distributor and micro-end-user respectively.	Pseudo-surrogate

Table 6

Also, figure 75 below shows the hierarchy of the business distributional phases described above. The grouping of the consumers will be done according to the following,

1. Specific end-users: make specific orders to be delivered directly to them for purchasing.
2. General end-users: buy what the manufacturer generally provides.

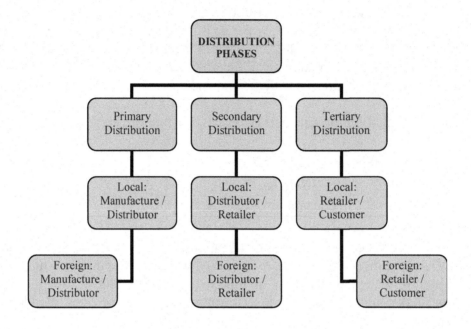

Figure 75. A typical business distribution phases.

The sets or groups shown in table 7 will be used to identify each region in a four-set Venn diagram representing a comprehensive business distribution activity.

Since no direct interaction exists between the two local and foreign distribution sets of D and D_s, the intersection $D \cap D_s$ which forms part of set notations (xv) and (xvi) represents a secondary distribution phase involving distributors and retailers. Similarly, the intersection $R \cap R_s$ which forms part of set notations (xiii) and (xiv) is representative of a secondary distribution phase involving retailers and consumers. Also, the intersections of (xxi) and (xxii) were defined based on the following. On the basis of hierarchy,

1. Distributors are on a higher level than the recipients they distribute products to.

2. Local distributions are higher activities compared to foreign distributions which are lower activities (see figure 75).

	GENERAL BUSINESS DISTRIBUTION ACTIVITY	
	Set Notation	**Description**
i.	D	Local manufacturer (local macro-distributor) equivalent to products.
ii.	R	Local distributor (local macro-end-user).
iii.	D^μ	Local retailer (local micro-distributor).
iv.	R^μ	Local consumer (local micro-end-user).
v.	D_s	Foreign manufacturer (foreign macro-distributor) equivalent to products.
vi.	R_s	Foreign distributor (foreign macro-end-user).
vii.	D_s^μ	Foreign retailer (foreign micro-distributor).
viii.	R_s^μ	Foreign consumer (foreign micro-end-user).
ix.	$D \cap R$	Mean quota for each member of R.
x.	$D_s \cap R_s$	Mean quota for each member of R_s.
xi.	$D_s \cap R$	Mean imported (borrowing) quota, -w for each member of R.
xii.	$D \cap R_s$	Mean exported (lending) quota, +w for each member of R_s.
xiii.	$D \cap R \cap R_s$	Mean quota for each member of R^μ.
xiv.	$D_s \cap R \cap R_s$	Mean quota for each member of R^μ.
xv.	$D \cap D_s \cap R$	Mean quota for each member of D^μ.
xvi.	$D \cap D_s \cap R_s$	Mean quota for each member of D_s^μ.
xvii.	Only D	Retained products after D distribute to R.
xviii.	Only D_s	Retained products after D_s distribute to R_s.
xix.	Only R	Retained products after R distribute to D^μ.
xx.	Only R_s	Retained products after R_s distribute to D_s^μ.
xxi.	$D \cap D_s$	Retained meso-products after D^μ distribute to R^μ.
xxii.	$R \cap R_s$	Retained meso-products after D_s^μ distribute to R_s^μ.
xxiii.	$D \cap R \cap D_s \cap R_s$	Mean quota for each specific local and foreign consumer in R^μ U R_s^μ via direct sale, indirect sales and e-commerce.
	+ w	Exported products by local manufacturer to 'foreign manufacturer' (Lending activity).
	- w	Imported products by local manufacturer from foreign manufacturer (Borrowing activity).

Table 7

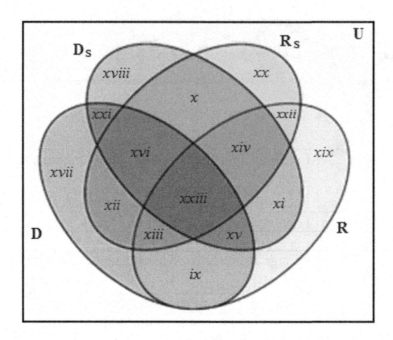

Figure 76. Identification of the activities of business distribution channel.

The Venn diagram shown in figure 76 identifies the corresponding positions of all the set notations for a business distribution channel activities.

Probability and Propensity Considerations

Let d be the number of distributum and r the number of recipients. Then by definition, the simplified probability ratio, P_o is given by (see chapter 2)

$$P_o = \frac{n(d \in D \cap R)/d}{n(D \cup R)/d} \cdot \frac{n(r \in D \cap R)/d}{n(D \cup R)/d} \quad if \ d < r$$

or

$$P_o = \frac{n(d \in D \cap R)/r}{n(D \cup R)/r} \cdot \frac{n(r \in D \cap R)/r}{n(D \cup R)/r} \quad if \ r < d.$$

The 'surrogate number', z is also given by

$$z = n(U_i) - n(D \cap R)$$

From chapters 2 and 3, the ideal active potential surrogate recipient number is by definition given as

$$n_i(\Phi_a) = \frac{n(D)}{n(U_i)} + \frac{n(D')}{n(U_i)} = 1$$

since

$$n(U_i) = n(D) + n(D').$$

The real active potential surrogate recipient number is defined as (see chapter 2)

$$n_r(\Phi_a) = \frac{n(D)}{n(U_r)}$$

where $n(U_i) = d + r$.

Hence, the propensity of the distribution is given, in terms of a ratio, as

$$\widetilde{P} = \frac{n_r(\Phi_a)}{n_i(\Phi_a)}$$

The simplified probability ratio and propensity for the local distribution phases will next be determined.

1. *Under Local Primary Distribution Phase*

Since there are more items than number of distribution channels, let the number of macro-distributor, D (i.e. manufactured products) be d and the number of macro-end user (i.e. distributor), R be r. Then from figure 77 which illustrates the situation

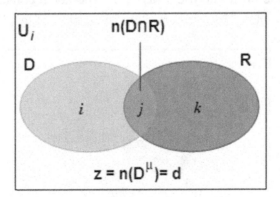

$$P_o = \frac{\dfrac{d}{r}}{n(D \cup R)/r} \cdot \frac{\dfrac{r}{r}}{n(D \cup R)/r} \quad where \; n(D \cup R) = i + r(j) + k$$

Figure 77

and i and k are retentions and j the mean quota. P_o is evaluated by first dividing each numerator and denominator by r. If R does not exist, the distributing group that would take over is the retailer group, D^μ. Hence, the surrogate number, z under local primary

181

distribution is, $z = n(D^\mu) = d^\mu$ where d^μ is the number of local micro-distributor (i.e. retailers). To find the propensity, we have

$$n_i(\Phi_a) = \frac{d}{n(U_i)} + \frac{r + d^\mu}{n(U_i)} = 1$$

since $n(U_i) = n(D) + n(D') = d + r + d^\mu$ and also

$$n_r(\Phi_a) = \frac{d}{n(U_r)} = \frac{d}{d + r}.$$

Therefore, the primary propensity is

$$\tilde{P}_p = \left(\frac{d}{d+r}\right) \Big/ \left(\frac{d}{n(U_i)} + \frac{r + d^\mu}{n(U_i)}\right) = \left(\frac{d}{d+r}\right)$$

2. *Under Local Secondary Distribution Phase*

Here, from figure 78 which illustrates the situation we can write

$$P_o = \frac{\dfrac{r}{d^\mu}}{i + d^\mu(j) + k} \cdot \frac{\dfrac{d^\mu}{d^\mu}}{i + d^\mu(j) + k}$$

$$\text{where } n\left[\left[(D_s \cup D)' \cap R\right] \cup (D_s \cap D)\right] = i + d^\mu(j) + k$$

If D^μ does not exist, the local micro-end-user), R^μ (i.e. local consumer) takes over as surrogate. Hence, $z = n(R^\mu) = r^\mu$ where r^μ is the number of local micro-end-user.

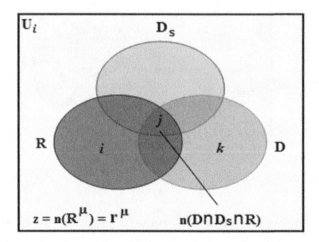

Figure 78

182

To find the propensity, we have

$$n_i(\Phi_a) = \frac{r}{n(U_i)} + \frac{d^\mu + r^\mu}{n(U_i)} = 1$$

and

$$n_r(\Phi_a) = \frac{r}{n(U_r)} = \frac{r}{r + d^\mu}$$

Therefore, the secondary propensity is

$$\tilde{P}_\Omega = \left(\frac{r}{r+d^\mu}\right)\bigg/\left(\frac{r}{n(U_i)} + \frac{d^\mu + r^\mu}{n(U_i)}\right) = \left(\frac{r}{r+d^\mu}\right)$$

3. *Under Local Tertiary Distribution Phase*

Here,

$$P_o = \frac{\dfrac{d^\mu}{r^\mu}}{i + r^\mu(j)} \cdot \frac{\dfrac{r^\mu}{r^\mu}}{i + r^\mu(j)}$$
$$\text{where } n(D_s \cap D) = i + r^\mu(j)$$

Since R^μ does not distribute but is only a terminal receiving end, its plausible surrogate, z cannot be determined from the Venn diagram depicting the scenario in figure 79. However, z can be computed using the formula quoted earlier on. Thus

$$z = n(U_i) - n(D_s \cap D \cap R)$$

which gives

$$z = n(U_i) - j \ .$$

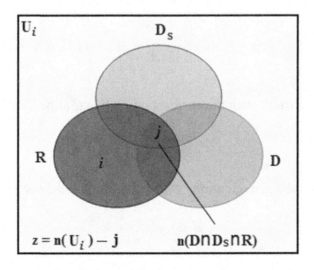

Figure 79

183

To find the propensity, we have

$$n_i(\Phi_a) = \frac{d^\mu}{n(U_i)} + \frac{r^\mu + n(U_i) - j}{n(U_i)} = 1$$

and

$$n_r(\Phi_a) = \frac{d^\mu}{n(U_r)} = \frac{d^\mu}{d^\mu + r^\mu}.$$

Therefore, the tertiary propensity, is given by

$$\tilde{P}_\tau = \left(\frac{d^\mu}{d^\mu + r^\mu}\right) \Bigg/ \left(\frac{d^\mu}{n(U_i)} + \frac{r^\mu + n(U_i) - j}{n(U_i)}\right) = \left(\frac{d^\mu}{d^\mu + r^\mu}\right)$$

The same procedure used for the local situations can be used to determine the simplified probability ratios and propensities for foreign scenarios.

Universal Linear Distributions

The nature of the processes involved in distribution channels turnout to be consistent throughout all space and time. It is this universality (or background independence) that will be scrutinized and established.

Consider a distribution involving D and R. Then by definition, the propensity will be given by

$$\tilde{P} = \frac{d}{n(U_r)} = \frac{d}{r(j+1)} \quad where \ n(U_r) = d + r = rj + r = r(j+1).$$

This equation can be written as

$$\frac{d}{r} = \tilde{P}(j+1) = \tilde{P}j + \tilde{P}$$

If we let

$$A_\rho = \frac{d}{r}$$

be the '**allocation statistic**', then the corresponding linear allocation distribution equation can be written as

$$A_\rho = \tilde{P}_{all} j + \tilde{P}_{all}$$

On the other hand, to analyze retentions in the distribution channels we let

$$d \equiv i \ and \ r \equiv k.$$

Then if we let

$$D_\rho = \frac{i}{k}$$

represent the '**retention statistic**', the corresponding linear retention distribution equation is defined as

$$D_\rho = \widetilde{P}_{ret} j + \widetilde{P}_{ret}$$

Since distribution channels consist of is a chain of intermediaries with each passing products down the chain, it is process-oriented. As such it occurs through time and space and so it is a flow concept. It can therefore be measured as a 'rate of sharing per period of time'. So, the retention and allocation statistics are essentially '**rate of retention**' and '**rate of allocation**' respectively. During an intra-phase analysis (specific) only one phase is analyzed. For example, values for the retention ratio i/k and j of the primary distribution phase the will be collected over time and analyzed. Also, for an inter-phases analysis (general) all the phases of distribution are analyzed simultaneously. Here, data will be collected for the retention ratio i/k and j for each

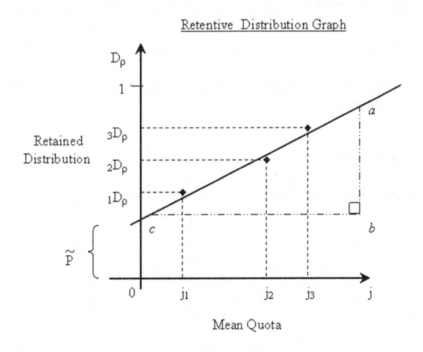

Figure 80. A linear retention graph.

phase per time. A plot of i/k against j will result in a specific and general linear retention graphs respectively. On the other hand, if the allocation ratio d/r and j values are collected for both intra-phase and inter-phases analysis the resulting plots will give specific and general linear allocation graphs respectively. Figure 80 shows a typical retention graph. From the graph, the slope is given by

$$Slope = y - intercept = \frac{ab}{bc} = \widetilde{P}$$

185

Since the graph is linear, it implies that the characteristics of the customers are directly proportional to those of the other recipients up the chain of distribution. Hence the following general inference for customers' purchasing behavioral pattern is deduced as shown in table 8.

Propensity $(\widetilde{P})$	Specific / General Mean Quota $(j$ or $\bar{j})$	Implications of Distribution Channels' Trending	Stability of Purchases
Large	Small	Customers have high natural tendency to buy less.	The lower purchases are stable.
Small	Large	Customers have low natural tendency to buy more.	The higher purchases are not stable.
Large	Large	Customers have high natural tendency to buy more.	The higher purchases are stable.
Small	Small	Customers have low natural tendency to buy less.	The lower purchases are unstable.

Table 8. General trending analysis of distribution channels.

In a similar manner as the retention analysis, the allocation statistic, A_ρ can be analyzed as was done for the retention statistic, D_ρ.

Marginal Distribution

From the linear retention and linear allocation equations,

$$\widetilde{P}_{all} = \frac{\left(\frac{d}{r}\right)}{(j+1)}$$

and

$$\widetilde{P}_{ret} = \frac{\left(\frac{i}{k}\right)}{(j+1)}$$

Observe that the mean quota of the distributions, j is increased by one unit. The extra distribution that would take place due to the additional unit of mean quota (assuming all other factors are fixed) is dubbed '**marginal distribution**', MD. By definition,

$$MD = \frac{\left(\frac{d}{r}\right) - \left(\frac{i}{k}\right)}{(j+1)} = \Delta\widetilde{P}$$

where $\Delta\widetilde{P}$ is the change in propensity.

186

A graph of distribution rate against distribution quota depicted in figure 81 shows the relationship between the linear retention and that of linear allocation curves. From the graph,

$$\Delta \widetilde{P} = \widetilde{P}_{all} - \widetilde{P}_{ret}$$

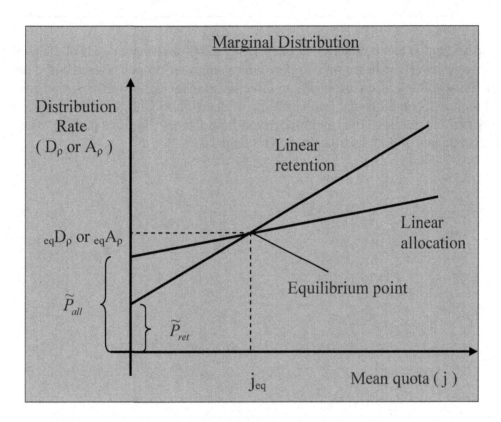

Figure 81. Graph of marginal distribution

Linear retention sharing (redundant-distribution analysis) involves hypo and hyper-quota. Therefore it invokes both inflationary and deflationary distributions. But the linear allocation sharing (desired-distribution analysis) involves an unbiased quota, and so promotes balanced distributions. The '**universal linear equilibrium point**' $(j_{eq},\,_{eq}D_\rho)$ or $(j_{eq},\,_{eq}A_\rho)$ is where the distributed rate of both linear retention and linear allocation are equal. This is the point where normalization of hypo or hyper-quota in the linear retention occurs.

Marginal distribution is, by definition, given by the change in the slope of linear allocation and linear retention curves. Since the distribution channel is fed by products produced by the manufacturer, there seem reason to expect a direct relationship between products and their distribution. In economics, marginal product is known to drive changes in the average product. A falling average physical product is known to imply that the marginal physical product must be less than the average and vice versa. In a similar fashion, it can be stated that,

A marginal distribution that is negative in value would imply the average distribution is falling. On the other hand, when the marginal distribution is positive it would imply that the average distribution is rising.

To wrap up here, it can be stated that the pricing and promotional needs of the marketing mix can definitively be based on computed intra and inter-propensities and also on marginal distribution. Consequently, an effective management of distribution channels should require the making and implementing of decision(s) with the aid of intra and inter-propensities and marginal distribution derived from the distributions passing through the physical channels of both local and foreign scenarios.

CHAPTER 8

DISTRIBUTIONAL COMPLEXITY

The application of the concepts of inflationary and deflationary distributions to a distributional complexity class comprised of a set of distributive sets of related resource-based complexity will be investigated.

DEFINITION OF P AND NP CLASSES

A complexity class by definition is a set of problems having related resource-based complexity. In dealing with resources required during computation, two complexity classes come into play. Decision problems that can be solved on a deterministic sequential machine in an amount of time that is polynomial in the size of the input are called class P. However, decision problems whose positive solutions can be verified given the appropriate information called the certificate in polynomial time are referred to as class NP. [18]

Alternatively, if an algorithm such as a Turing machine or Pascal program with unbound memory is able to give correct answer for any input string of length n in at most n^k steps, where k is some constant independent of the input string, then the problem can be solved in polynomial time and it is referred to as class of P. On the other hand, suppose an algorithm, say Y (d,c) takes two arguments namely,

1. An input string d to a decision problem and

2. A string c representing a proposed certificate.

Let Y produces a Yes or No answer in at most n^k steps (where n is the length of d and k does not depend on d). Assuming d is a Yes instance of the decision problem if and only if there exists c such that Y (d,c) returns Yes, it implies the problem can be solved in non-deterministic polynomial time. Hence it is said to be in the NP class.

Problem Definition

In computational theory, use is made of algorithms with various restrictions called certificates to study which languages are decidable. The problems that are dealt with are those called decision problems. These are problems that can be answered with 'Yes' or 'No'. In other words, a decision problem takes as input some string (i.e. a set of finite-lengthy questions) and requires as output either 'Yes' or 'No'. When decision problems are encoded (expressible in any base greater than 1) the result is a set of strings called language, which invariably represent the instances (i.e. different occurrences of the same problem) of the original decision problem.

The monolithic question at hand now is whether or not P is equal to NP. This question is basically asking if 'Yes'- answers to a 'Yes'-or-'No'-question is quickly verifiable can the said answers themselves also be computed quickly within a polynomial time? With restrictions to 'Yes' or 'No' problems found to be unimportant, the idea of allowing more complicated answers lead to the function problems version of the decision problem. It is similar to the computation/verification relationship between P and NP. The complexity class FP is by definition the set of function problems solvable by deterministic Turing machine in a polynomial time. The function problem version of the decision problem class P is FP. In other words, FP = FNP is the proven equivalence of P = NP. [19] Also, it has been shown that FP = FNP if and only if P = NP.

SOLVING THE PROBLEM

Using a common equation based on the general equations of the inflationary and deflationary distributions via the law of equity distribution, we can write

$$X = zY \pm W \qquad (46)$$

Let the above generic equation define any decision problem. Then X is the solution set (e.g. distributum), zY the problem set (e.g. normalized inflated or deflated quota) and W (e.g. surrogated lending or borrowing) the restriction set or the certificate. The problem set has a special rule for generating its members. This means it is ordered and as such Y represents the 'shortest program' or complexity of the set. [20] Also, z (e.g. inflated and deflated quota which are always whole number in value) will represent the number of repetitions of the 'shortest program'. This implies the problem set is ordered and as such not random because the complexity is smaller or shorter than the problem set. Consequently, there exist mathematical formulations or laws within the problem set which has shorter complexity. Thus, for a general non-randomize case equation (46) is applied while a general randomize case is given as

$$X = Y \pm W$$

Note that the randomize situation is a very complex case because the set zY is lesser in complexity than the set Y (e.g. interaction between recipients and surrogate recipients). In other words, the set zY is a 'simple set' and that of Y is a 'complicated set'.

A Rule for Determining the Ease of Checking a Problem

Let the number of repetitions in the problem set z be represented by z_k which is the normal number of repetitions (i.e. unbiased quota) of a simplified problem, z^+ the inflated number of repetitions of a simplified problem and z^- the deflated number of repetitions of a simplified problem. Then an inflationary problem's number range will be defined as $[0, z^+]$ which is the whole numbers from 0 to z^+. Remember that the least whole number that the unbiased quota can represent is 0. Similarly, a deflationary problem's number range will be defined as $[0, z^-]$. Note that the number of repetitions in the problem is the solution sort here in this hypothetical distributional system.

By definition,

> *If z_k belongs to the set of range numbers for a particular distribution case, then the distribution case in question is determinable or easy to check if solution exists else it can not be easily checked if solution exists.*

As an example, let the following be arbitrary values of z: $z_k = 3$, $z^+ = 5$, $z^- = 2$. Then for the deflationary case (i.e. $z^- = 2$), the value of z_k, indicated by the small empty circle cannot be found within the deterministic range $[0, z^-]$ as depicted in figure 82.

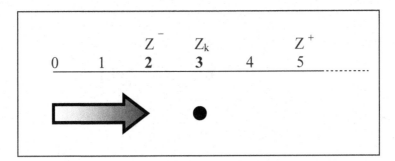

Figure 82.

For the inflationary case (i.e. $z^+ = 5$), the value of z_k does lie within the range as depicted in figure 83 below.

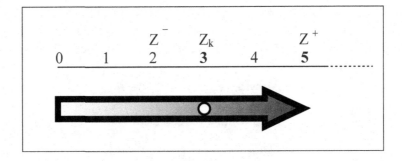

Figure 83.

191

Thus, since by definition

$$z^- < z_k < z^+$$

it can generally be concluded that,

> **_Inflationary distribution cases are easy to check while on the contrary the deflationary distribution cases cannot be easily checked if solution exists._**

Generalized Decision Problem

With the prevalence of distribution phenomenon common place in nature, its universality cannot be overemphasized. It permeates area such as mathematics, science and technology among others. Without distributive processes, nature is bound to fizzle out of existence. Consequently, the equation of the law of equity distribution will be used as the basis for defining a general decision problem.

For a normal distribution case,

$$X = z_k Y + W_k$$

For abnormal distribution cases (i.e. inflationary and deflationary cases),

$$X = (z_k \pm \Delta z)Y + (W_k \pm \Delta W)$$

given that

$$z_k + \Delta z = z^+, \qquad z_k - \Delta z = z^-$$
$$W_k + \Delta W = W^+, \qquad W_k - \Delta W = W^-$$

and also

$$z_k \pm \Delta z = z, \quad and \quad W_k \pm \Delta W = \pm W$$

where W_k is the residual set, ΔW the change in the restricted set or certificate, W^+ the inclusion set, W^- the exclusion set and W the composite restriction set.

Uniquely, the different abnormal distribution cases can be expressed as follows,

$$Normal \qquad : X = z_k Y + W_k$$
$$Inflationary : X = z^+ Y - W$$
$$Deflationary : X = z^- Y + W$$

192

Generally, a critical analysis shows that for

Inflationary decision questions: $|z^+| > |W|$ (*Checkable*)

Deflationary decision questions: $|z^-| < |W|$ (*Non – checkable*)

where $|z^+|$ is the modulus of z^+, $|z^-|$ the modulus of z^- and $|W|$ the modulus of W.

Relationship between P and NP

The inflationary equation,

$$X = z^+Y - W$$

where $z^+ = - W + 1$, $z_k + 2$, $z_k + 3$, … represents the infinite possibilities of a simplified problem. This will shortly be used to help establish the relationship that exists between P and NP.

Let X represent a set with a language representing the class P, then $(z^+Y- W)$ will be a set with a language representing the class NP because z^+ has infinite possibilities. It is the infinite possibilities of z^+ that makes it 'non-deterministic'. Hence, it could be stated that,

$$P = NP$$

in accordance with the inflationary equation.

Formal Definition of Reduction

Let A and B be two languages. Then A is reducible to B, denoted A < B, if and only if there exists a function f such that,

1. For all x in A, $f(x)$ is computable in deterministic polynomial time

2. For all x, x is in A if and only if $f(x)$ is in B

By definition, 'Karp-reduction' exists if there is a deterministic polynomial time function that maps 'yes'- instances in A to 'yes'-instances in B and 'no'- instances in A to 'no'-instances in B. [21] In such a case, A is reducible to B or 'A is Karp-reducible to B'. Again, if A< B, it could be said that A is 'no harder' than B.

Formal Definition of NP-Completeness

The concept of NP-completeness is important in that it is informally thought that they are the ones most likely not to be in P. Consequently, if a single NP-complete problem could be shown to be in P, it automatically follows that P = NP.

A language L is NP-complete if and only if

1. L is in NP.
2. For any language L′ in NP, L′ < L. [21]

In other words,

1. If a language is in NP then it is NP-complete.
2. Every other language in NP must be no harder than it.

Let us exemplify the inflationary equation

$$X = z^+Y - W$$

by using the ratio of 5/3. Since 5/3 = 1 remainder 2, x = 5, y = 3 and z_k = 1. Let

$$z^+ = 4$$

then P's language L (in base ten) is 5. This implies NP's language is also 5 according to the relationship established between P and NP earlier on. It further implies that L is in NP. The language of y (which is equal to 3) say L′ is in NP and L′ < L which means L′ is reducible to L.

Proof of NP-Completeness

Let the instance of a given language run from zero to the language and let f(x) be a function given arbitrarily by say,

$$x = f(y) = y + 2$$

where y is any member of the set L′ or set Y and x is any member of the set L or set X. A mapping can be established between L′ and L as depicted in figure 84.

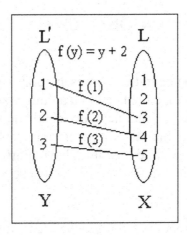

Figure 84. Mapping between L and L´ based on the function f(y).

In proving reduction, it is noted that

1. For all y in L´, f(y) is computable in deterministic polynomial time.

2. For all y, y is in L´ if and only if f(y) is in L.

Hence, L´ is Karp-reducible to L. As such L´ is no harder than L and furthermore L is NP-complete and so is L´.

CONCLUSION

It is intriguing how concise mathematical representations of the phenomenon of equity distributions (sharing processes), elucidated in a uniquely simple but realistic manner the scenarios involving the distributor, recipient, dividend and surrogate sets. Consequently, this led to a comprehensive degeneration of indeterminate(s) and related paradoxical outcome(s). Hitherto, such undesired outcomes have blurred the dynamism and/or clogged the arteries of concise clarity that nourished the very description of the structural body of the universe of mathematics.

The law of equity distribution facilitated reproducible results of critical mathematical conundrums such as division by zero and the ratio of zero to zero. It also effectively identified two sets of numbers namely the non-distributive and distributive numbers. While distributive numbers form real number field, the non-distributive numbers form an abstract number field which is patently governed by a set of abstract field axioms. Further, under equity distribution, the concept of redistribution in the real world was shown to lack adequate natural tendency to flourish. For it to significantly occur even though it will eventually fail, an external force has to be introduced thus making it unpopular. However, the idea of surrogation among its tenet showed that the absence of a recipient group naturally facilitates at least a single recipient with a mean propensity.

On the value of pi deduced through mapping-intersection argument, a simple but significant fundamental question begs to be answered. Is pi really equal to 3? The Hebrews, as it is known, used the integer 3 as the value of pi. [15] On the contrary, pi has been accepted as irrational. [22] It is also known to be transcendental in nature. [23] Using Hitachi SR8000 supercomputer, the determined modern irrational value of pi (to more than one trillion decimal digits) is given as 3.14159265.... [24] Usually, the considered standard approach to the rigorous foundation of Calculus is the mathematical concept of limits. It rests on the rigorous development of Calculus from precise axioms and definitions. The most recent approaches that have certainly been used to determine the irrational value of pi are mathematical limits and series. [25] Since the said modern irrational-pi results from the concept of mathematical limits (as one of its approaches), it would be definitely easier to re-investigate the appropriateness of the limits concept in terms of exactitude. Perhaps, if it can be proved that the concept of limits is not truly exact, then the integer-pi must be exact. If this turns out to be the case, it would be a good suggestion that the irrational-pi be considered as an excellent approximation and symbolized as $\tilde{\pi}$ while the symbol for the exact integer-pi is made $\bar{\pi}$. Also, one of the other approaches that can facilitate the above investigation would be the use of Euclidean geometry to establish if pi is an integer or irrational. Whatever the outcome of this essential endeavour might be, one thing is for sure crystal clear. It will certainly settle all the plethora of fundamental questions about and/or linked to pi.

The assertion by Cantor's continuum hypothesis came within the scope of this literature. Simply put, the hypothesis states that there exists no infinite set with a cardinality that can fit between the infinity of spatial points called aleph-null (or continuum) and the infinite set of integers called aleph-one. Some years later, both Gödel and Cohen proved independently and differently that the continuum hypothesis cannot be proven or disproved by standard axioms of contemporary set theory. In other words, the continuum hypothesis came to exemplify a major mathematical conundrum that definitively cannot be solved by using universally accepted system of axioms on which most mathematics are built. To marinade this problem with other complexities, the lack of consistency in axioms and rules nullifies possible answers to finding all that lies beyond the aleph-one (i.e. the set of integers). Once again, some mathematicians have chosen, as a means of comfort and not positive rigorous solution, to assume that the continuum hypothesis is wrong! But analysis done within the conceptual framework of '*The Obelus Set Theory of Equity Distribution*' proved otherwise. It showed that a single geometrical point contains infinitely 'aleph-half' geometrically latent points. As it turned out, the transfinite 'aleph-half' lies between aleph-null and aleph-one, hence its name. The merits of a wholistic approach to the process of sharing, once more stands out here as a veritable means to unravel serious mathematical conundrums of both basic and complex nature.

The provision of a set definition for a geometrical point helps to settle concisely the non-rigorous nature of the intuitive way in which its characteristics have hitherto been defined. It puts forth and elucidates the much needed concepts of a geometrical point and for that matter a line and a plane, upon which all other geometric definitions and concepts are built on. This should help remove any vagaries surrounding the undefined concepts of a geometrical point. The central role played by the distribution of geometrical points along the circumference of a circle which passes through the points corresponding to its diameter is pivotal to the much needed concepts made known here.

On hyper and hypo-distributions, the role of surrogation evinced how both distributum and distribient are mutually exclusive under inflationary distributive processes and mutually inclusive under deflationary distributive processes respectively. The characteristic resonance nature of surrogation in both hyper and hypo-distributions facilitates distributional stabilization through surrogation dynamics. The analysis of distributional retentions and allocations are made possible. With the concept of marginal distribution, the rise or fall of the average distribution is measurable and so more controlled. Not to mention other areas, measures of the said distributional statistics are bound to be most helpful in the management of business distribution channels and the market mix as a whole.

The dichotomy between the abstraction of the real world into the mathematical world of abstract relationships and the specialization of the mathematical world into specificity of things sincerely begs the following pertinent question: Can any part or whole of the mathematical world of abstract relationships exist without any instantiation in the real world? A classic example of such an important question is the so-called P class (which involves problems solvable by conventional computers in time polynomial to the input size) and NP class (which requires one to find a solution of a problem where it is feasible to quickly verify that the solution is correct) problem that challenges the core of theoretical computer science and contemporary mathematics to date. The monolithic question to this P-NP problem is if P is equal to NP which simply put implies: Can the ability to recognize an answer translate into the ability to find an answer? To some, this represents the nature of mathematical thought itself. If P should be equal to NP, some of the stunning practical consequences would include,

1. The solution for many optimization problems.

2. The transformation of mathematics via the use of computers to find formal proof to any theorem with a reasonably length proof.

3. No fundamental difference between solving a problem and recognizing the solution of the problem.

Scott Aaronson of Massachusetts Institute of Technology soberly put it this way, that anyone who could appreciate a symphony would be Mozart or follow a step-by-step argument would be Gauss or recognize a good investment strategy would be Warren Buffett. Why cannot this be true? P being equal to NP only confirms the universality of '**intelligent adaptation**'. In human physiology, the loss of one's sense of sight is compensated for by the strengthening of other senses such as smell, hearing and/or touch. Here, the extra-sensitization of any of the working senses permits the verification of the solution of the 'sight problem' to allow the blind victim to 'see'. This is equivalent to that derived from the sense of sight which is by design able to compute the solution of a 'sight problem'. Whereas sensory organs respectively have different anatomy and physiology, the paramount question here is: who 'sees'? Put another way; is the realization of an image via intelligent adaptation less valuable than that through the eye? It is certainly not true, since all perceptions including sight are actually attained in the brain and not in the sensory organs naturally or adaptively associated with them. In essence, irrespective of the route to the solution of the said 'sight problem' they are all attainable just as when needed. As far as the consequences go, they evince that the real world is universally flavoured with equity distribution. While many currently think that P is not likely to be equal to NP, the 2007 Gödel Prize for outstanding journal articles in theoretical computer science awarded paper, "Natural Proofs" indicated that unless widely held conjectures are violated a wide class of proof techniques cannot be used to resolve the P-NP and related problems. [26] From the basis of the law of equity distribution, it was possible to

1. Demonstrate the ease of checking a problem via inflationary and deflationary distributions.

2. Generalized the decision problem and show the equality between P and NP.

3. Provide a proof for NP-completeness via inflationary distributions analysis.

In this book, the versatility of *Obelus* set theory has been demonstrated in order to encourage others to delve into it and make use of it with the view of providing wholistic solution(s) within the domain of mathematics, science and technology.

Finally, it is my hope that in the near future the elaboration and/or extrapolation of the principles set forth in this book will provide the necessary mathematical truism for, if not all, some existing seemingly insurmountable mathematical problems and those that are yet to be perceived.

BIBLIOGRAPHY

1. Derbyshire, J. Prime Obession: Bemhard Rieman and the Greatest Unsolved Problem in Mathematics. New York: Penguin, pp. 266, 2004.

2. Weisstein, Eric W. "Division by Zero." From MathWorld – A Wolfram Web Resource. http://mathworld.wolfram.com/DivisionbyZero.html

3. Dummit, D. S. and Foote, R. M. "Field Theory." Ch. 13 in *Abstract Algebra, 2nd ed.* Englewood Cliffs, NJ: Prentice-Hall, pp. 422-470, 1998.

   Ferreirós, J. "A New Fundamental Notion for Algebra: Fields." §3.2 in *Labyrinth of Thought: A History of Set Theory and Its Role in Modern Mathematics.* Basel, Switzerland: Birkhäuser, pp. 90-94, 1999.

   Nagell, T. "Moduls, Rings, and Fields." §6 in *Introduction to Number Theory.* New York: Wiley, pp. 19-21, 1951.

4. Knuth, D.E. "Two Notes on Notation." Amer. Math. Monthly 99, 403-422, 1992.

   Knuth, D.E. The Art of computer Programming, Vol.1: Fundamental Algorithms, 3rd ed. Reading, MA: Addison-Wesley, p.57, 1997.

5. Wells, D. *The Penguin Dictionary of Curious and Interesting Numbers.* Middlesex, England: Penguin Books, pp. 23-26, 1986.

6. Http://en.wikipedia.org/wiki/Indeterminate_form

7. Http://en.wikipedia.org/wiki/Defined_and_undefined#Zero_to_the_zero_power

8. Graham, Ronald, Knuth, Donald and Patashnik, Oren, Concrete mathematics: A Foundation For Computer Science, Addison-Wesley Publishing Co., Reading, Mass, 2nd edition: January 1994.

9. Gödel, K., 1940, 'The Consistency of the Axiom of Choice and the Generalized Continuum Hypothesis', Ann. Math. Studies, 3.

10. Gamov, George, One Two Three …Infinity, Dover Publications, Inc, Mineola, New York. p. 20, 1988.

11. 'Cardinality of the Continuum', http://en.wikipedia.org/wiki/Infinity.

12. http://www.suitcaseofdreams.net/Infinity_Parodox.htm

13. Casey, J. "The Point." Ch. 1 in A Treatise on the Analytical Geometry of the Point, Line, Circle, and Conic Sections, Containing an Account of Its Most Recent Extensions, with Numerous Examples, 2nd ed., rev. enl. Dublin: Hodges, Figgis, & Co., pp. 1-29, 1893.

14. www.learner.org/courses/learningmath/geometry/keyterms.html

15. Wells, D. *The Penguin Dictionary of Curious and Interesting Numbers.* Middlesex, England: Penguin Books, pp. 48-55 and 76, 1986.

16. Holliday, David and Resnick, Robert. Physics, John Wisley & sons (New York), 1978, 3rd. Ed. p. 424.

17. Venn, John. *On the diagrammatic and mechanical representation of propositions and reasonings,* The London, Edinburg, and Dublin Philosophical magazine and Journal of Science, 9 (1880), p. 1-18.

18. Sipser, Michael: *Introduction to the Theory of Computation, Second Edition, International Edition,* page 270. Thomson Course Technology, 2006. Definition 7.19 and Theorem 7.20.

19. Bellare, M. and Goldwasser, S. The complexity of decision versus search. SIAM Journal on Computing, February 1994, Vol. 23, No. 1.

20. Barrow, John D., Pi in the Sky, Little, Brown & company (Boston) 1992. First Edition, p. 138.

21. Garey, M.R. and Johnson, D. S. Computers and Intractability, W.H. Freeman, 1979.

22. Lindemann, F. "Über die Zahl π." *Math. Ann.* 20, 213-225, 1882.

23. Kanada, Y. "Sample Digits for Decimal Digits of Pi." Jan. 18, 2003. http://www.super-computing.org/pi-decimal_current.html.

    Peterson, I. "MathTrek: A Trillion Pieces of Pi." Dec. 14, 2002. http://www.sciencenews.org/20021214/mathtrek.asp.

24. 'Computing π', http://en.wikipedia.org/wiki/Pi.

25. Nagell, T. "Irrationality of the numbers e and π." §13 in *Introduction to Number Theory.* New York: Wiley, pp. 38-40, 1951.

    Niven, I. M. *Irrational Numbers.* New York: Wiley, 1956.

Struik, D. *A Source Book in Mathematics, 1200-1800.* Cambridge, MA: Harvard University Press, 1969.

Borwein, J. and Bailey, D. *Mathematics by Experiment: Plausible Reasoning in the 21st Century.* Wellesley, MA: A K Peters, 2003.

Lambert, J. H. "Mémoire sur quelques propriétés remarquables des quantités transcendantes circulaires et logarithmiques." *Mémoires de l'Academie des sciences de Berlin* **17**, 265-322, 1761.

Hermite, C. "Sur quelques approximations algébriques." *J. reine angew. Math.* **76**, 342-344, 1873. Reprinted in *Oeuvres complètes, Tome III.* Paris: Hermann, pp. 146-149, 1912.

Schröder, E. M. "Zur Irrationalität von π^2 und π." *Mitt. Math. Ges. Hamburg* 13, 249, 1993.

Stevens, J. "Zur Irrationalität von π." *Mitt. Math. Ges. Hamburg* 18, 151-158, 1999.

26. Razborov, A. A. and Rudich, S. "Natural Proofs". *Journal of Computer and System Sciences* (1997). **55**: 24–35.

INDEX

202

mapping period, 145, 148, 149, 151
marginal distribution, 186, 188, 197
marketing mix, 176, 188
mean recipient set, 77, 78
mean zero, 77, 90, 91, 93, 118, 157, 158, 159
Meso-distributor, 177
meso-end-user, 177
Micro-distributor, 177
micro-end-user, 177, 179, 182
Mono-Polar
    Division, 26
multiple-surrogate, 64

N

negative integer sigma bond, 118
negative nuclear integer, 118
negative sigma integer, 118
negative SNDZ, 91, 92
negative zero, 51, 87, 88, 90, 92, 94, 118, 121, 123
net cardinality, 141, 151
neutral number, 98
neutral number group, 98
neutral unit integer, 89, 90, 109, 119
neutral unit number, 99, 100, 103
neutralization process, 157
non-distributive cardinality, 100
Non-Distributive Number', 17
non-distributive zero', 18
Non-Polar
    Division, 28
Non-Zero
    Division, 22, 29
NP-completeness, 194
Null
    Division, 10, 24, 29, 46, 47
null power set, 141, 143, 144, 150
Null Property, 10
nullified integers, 118
numeric atoms, 118
numeric di-polarity, 98
numeric induction, 98
numeric mono-polarity, 98
numeric pi bonds, 119
numeric sigma bond, 118
numerical
    arithmetic, 2, 3, 5, 11, 21, 24

O

obelus space, 85

P

passive null set, 39
passive surrogate, 52, 54
periodic quotient, 76, 77
pi, 138
pi distributions, 119
point continuum, 156
point singleton, 156
point-density factor, 151

Polarization, 121, 122
polynomial time, 189, 190, 193, 195
positive integer atom, 118
positive nuclear integer, 118, 119
positive sigma integers, 118
positive SNDZ, 91, 92
positive zero, 87, 88, 89, 92, 93, 94, 118
primary distribution, 177, 182, 185
primary propensity, 182
probability, vi, 30, 31, 42, 43, 44, 45, 46, 66, 67
process-oriented, 185
Product-Difference Analysis, 73
Product-Sum Analysis, 72
propensity, vi, 42, 46, 47, 48, 49, 50, 51, 52, 55, 56, 67, 68, 78, 196

Q

quota, 4, 6, 138, 144
quotient probability, 31, 32, 56, 67, 68

R

rate of allocation, 185
rate of retention, 185
receivables per number of group, 30
reciprocity, 7, 8, 10, 148, 149
redistribution, vi, 55, 196
reduction principle, 151
Redundant or Degenerated
    Division, 24, 29
resonance equilibrium, 164, 166, 168, 169, 170, 171, 173, 174
retention
    bracket, 2, 3, 11, 13, 162
retention statistic, 185, 186

S

secondary distribution, 177, 178
secondary propensity, 183
semi-reduction process, 150
set resonance, 78
set surrogate resonance, 78
sigma distributions, 118
simplified probability, 180, 181, 184
single-geometrical-point, 154
Specific end-users, 177
standing set surrogate resonance, 164, 166
surrogate number, 32, 38, 56, 68
surrogate quotient, 32
surrogate recipient, 32, 37, 38, 41, 52, 54, 75
surrogate wave beat, 136
surrogate waves, 76, 77, 96, 121, 125, 128
surrogation, 47, 49, 52, 68

T

tertiary distribution, 177
tertiary propensity, 184

U

unbiased quota, 163, 164, 169, 172, 176, 191
undefinededness, 28
unit donor distribution, 75
unit negative pi integer, 119
unit negative type integer, 118
unit positive pi integer, 118, 119
unit positive type integer, 118
Unit Property, 10
universal zeroth law, 88
unsigned zero, 77, 121, 123, 124, 157

V

void zero, 93, 118, 120

Z

zero induction, 121
Zero-Dividend
    Division, 23
zeroth arithmetic, 158
zeroth dipole, 159
zeroth dipole pairs, 159
zeroth dipoles, 159
zeroth even number, 88, 91
zeroth induction, 159
zeroth odd number, 88, 91, 93
zeroth polarization, 159, 160
zeroth power singleton, 159
zeroth set, 158
Zeroth-Exclusive, 9
Zeroth-Inclusive, 9
Zilch
    Division, 23